**Springer Theses**

Recognizing Outstanding Ph.D. Research

## Aims and Scope

The series "Springer Theses" brings together a selection of the very best Ph.D. theses from around the world and across the physical sciences. Nominated and endorsed by two recognized specialists, each published volume has been selected for its scientific excellence and the high impact of its contents for the pertinent field of research. For greater accessibility to non-specialists, the published versions include an extended introduction, as well as a foreword by the student's supervisor explaining the special relevance of the work for the field. As a whole, the series will provide a valuable resource both for newcomers to the research fields described, and for other scientists seeking detailed background information on special questions. Finally, it provides an accredited documentation of the valuable contributions made by today's younger generation of scientists.

## Theses may be nominated for publication in this series by heads of department at internationally leading universities or institutes and should fulfill all of the following criteria

- They must be written in good English.
- The topic should fall within the confines of Chemistry, Physics, Earth Sciences, Engineering and related interdisciplinary fields such as Materials, Nanoscience, Chemical Engineering, Complex Systems and Biophysics.
- The work reported in the thesis must represent a significant scientific advance.
- If the thesis includes previously published material, permission to reproduce this must be gained from the respective copyright holder (a maximum 30% of the thesis should be a verbatim reproduction from the author's previous publications).
- They must have been examined and passed during the 12 months prior to nomination.
- Each thesis should include a foreword by the supervisor outlining the significance of its content.
- The theses should have a clearly defined structure including an introduction accessible to new PhD students and scientists not expert in the relevant field.

Indexed by zbMATH.

More information about this series at http://www.springer.com/series/8790

Cecilia Tosciri

# Higgs Boson Decays into a Pair of Bottom Quarks

## Observation with the ATLAS Detector and Machine Learning Applications

Doctoral Thesis accepted by
University of Oxford, Oxford, UK

Springer

*Author*
Dr. Cecilia Tosciri
Department of Physics
University of Chicago
Chicago, IL, USA

*Supervisor*
Prof. Daniela Bortoletto
Particle Physics Department
University of Oxford
Oxford, UK

ISSN 2190-5053 ISSN 2190-5061 (electronic)
Springer Theses
ISBN 978-3-030-87937-2 ISBN 978-3-030-87938-9 (eBook)
https://doi.org/10.1007/978-3-030-87938-9



This Springer imprint is published by the registered company Springer Nature Switzerland AG
The registered company address is: Gewerbestrasse 11, 6330 Cham, Switzerland

*To Marco*

# Supervisor's Foreword

The discovery of the Higgs boson and associated Higgs field in 2012 was a triumph for experimental and theoretical particle physics and marked a landmark step in the understanding of our universe. The Higgs is an essential component of the Standard Model of fundamental particles and interactions (SM). Through spontaneous symmetry breaking, the Higgs field gives masses to the $W$ and $Z$ bosons. Yukawa couplings, which describe the interaction between the scalar Higgs field and massless fermion fields, generate the masses of quarks and charged leptons. Since the masses of fermions are very different, Yukawa couplings must vary dramatically in magnitude. The SM does not explain the size of the Yukawa couplings, which are just parameters to be measured.

The observation of the Higgs boson decaying into pairs of $\tau$ leptons provided the first evidence of Yukawa interactions. But only in 2018, 6 years after its discovery, did we finally conclusively observe the favoured decay of the Higgs boson into a pair of $b$ quarks ($H \rightarrow b\bar{b}$), which according to the SM, accounts for almost 60% of all possible Higgs decays. Observing this decay mode and measuring its rate was a critical step in confirming the mass generation for fermions via Yukawa interactions, as predicted in the SM.

In her thesis, Dr. Cecilia Tosciri concisely introduces the SM of particle physics, the ATLAS experiment, and machine learning. Her contributions to the observation of Higgs boson decays to a pair of $b$ quarks, one of the most demanding analyses carried out at the LHC, are presented in detail. Since the LHC collisions produce $b$-quark pairs in great abundance, it is essential to select events where the Higgs boson appears alongside a $W$ or $Z$ particle, making the events easier to tag. The analysis relies on machine learning to enhance signal sensitivity and reduce the impact of background processes.

Dr. Tosciri also used the data to measure the simplified fiducial cross sections in $VH(H \rightarrow bb)$ as a function of the vector boson momentum and extract, for the first time, effective field theory (EFT) parameters that can be accessed through this process.

Dr. Tosciri's D.Phil. was funded through participation in the AMVA4NewPhysics Horizon 2020 network, which focused on advanced statistical learning tools. She used a secondment at the Yandex and Higher School of Economics University Open Joint Laboratory, which was part of the AMVA4NewPhysics network, to develop a novel strategy for the fast simulation of the ATLAS forward calorimeter. The fast simulation is based on a frozen library approach, storing pre-simulated electromagnetic showers to model the detector response quickly. She collaborated with researchers at the Laboratory, who worked on similarity search techniques for search engines, and implemented this approach in ATLAS. Similarity search techniques are usually used in image feature matching in large-scale databases and semantic document retrieval. The approach is well suited for selecting electromagnetic showers from the frozen library and using these to model the detector response. The new method developed by Dr. Tosciri outperforms the standard fast simulation technique used in ATLAS.

I believe that both beginning students and experts will enjoy reading this thesis, which represents an important step towards the era of precision Higgs physics measurements at the LHC and effectively uses advanced machine learning methods.

Oxford, UK
May 2021

Prof. Daniela Bortoletto

# Abstract

The discovery in 2012 of the Higgs boson at the Large Hadron Collider (LHC) represents a milestone for the Standard Model (SM) of particle physics. Since then, most of the SM Higgs production and decay rates have been measured at the LHC with increased precision and found to be consistent with the predictions.

This thesis presents the analysis that led to the observation of the SM Higgs boson decay into pairs of bottom quarks. The analysis exploits the production of a Higgs boson associated with a vector boson whose signatures enable efficient triggering and powerful background reduction. The main strategy to maximise the signal sensitivity is based on a multivariate approach. The analysis is performed on a dataset corresponding to a luminosity of $79.8\,fb^{-1}$ collected by the ATLAS experiment during Run-2 at a centre-of-mass energy of 13 TeV. An excess of events over the expected background is found with an observed (expected) significance of 4.9 (4.3) standard deviation. A combination with results from other searches provides an observed (expected) significance of 5.4 (5.5). The corresponding ratio between the signal yield and the SM expectation is $1.01 \pm 0.12(\text{stat.})^{+0.16}_{-0.15}(\text{syst.})$.

The measurement of cross sections in exclusive regions of phase space of the production times the branching ratio is reported as well. These measurements are used to search for possible deviations from the SM with an effective field theory approach, based on anomalous couplings of the Higgs boson.

This thesis also describes a novel technique for the fast simulation of the forward calorimeter response, based on machine learning methods. The new simulation approach outperforms the default technique used by ATLAS.

# Preface

The work presented in this thesis is the result of a cooperation between myself and other members of the ATLAS Collaboration. The purpose of this manuscript is to present the main results achieved throughout this collaboration and describe my original contributions in the effort of producing them. Results and figures from works published by others have been clearly attributed in this thesis.

During my D.Phil., as a Marie Skłodowska-Curie fellow, I also collaborated with members of other institutions and collaborations, in the context of an interdisciplinary European Innovative Training Network (ITN) called *advanced multivariate analysis for new physics* (AMVA4NewPhysics). The major projects carried out within ATLAS and AMVA4NewPhysics during my D.Phil. are outlined in chronological order in what follows.

**Classification and regression studies for Higgs pair production in the resolved $b\bar{b}b\bar{b}$ final state** This phenomenological study, performed in collaboration with the AMVA4NewPhysics ITN, examined the potential of machine learning applications in the discrimination of Higgs pair production from background, and explored the regression technique as an additional tool for the improvement of the classification. For this project, I contributed to the Monte Carlo production of signal and background samples, I implemented the event selection for the $b\bar{b}b\bar{b}$ final state, and contributed to the implementation of the algorithms (*deep neural network* and *boosted decision tree*) necessary for the classification studies. The work is documented in *Deliverable D1.1* at Ref. [1]. In the second phase of the project, I have adapted the multivariate techniques and applied them to ATLAS simulated samples. Preliminary results are documented in Ref. [2]. The di-Higgs classification project is not covered in this thesis.

**Jet Energy Resolution (JER) studies for $b$-jets** Using the *direct balance* method in $\gamma$+ jets events, I have studied the jet response and JER for different b-tagging working points, and I have evaluated the systematic uncertainties on the JER. This work, documented in Ref. [3], qualified me as an author of the ATLAS Collaboration, since the beginning of 2018. The project is not discussed in this thesis, although the direct balance method is described in Sect. 5.5.2.

**Observation of Higgs boson decays into a pair of bottom quarks in the vector boson associated production channel** Since February 2018, I actively worked on the $VH$ $(H \to b\bar{b})$ analysis performed with Run-2 data, collected by the ATLAS experiment and corresponding to an integrated luminosity of $79.8 fb^{-1}$. Chapter 7 of this thesis discusses the main aspects of the analysis that led to the observation of $H \to b\bar{b}$ decay process. I contributed to the results with extensive studies on $b$-jet energy corrections, in particular evaluating the optimal working point for the application of the correction, outlined in Sect. 7.4.1. I documented the studies in the supporting note regarding the definition and selection of the analysis objects [4]. I also contributed as analyser of the channel, especially in the production of the samples needed for the training of the multivariate analysis (MVA) based on a boosted decision tree. Additionally, I was part of the team that performed the $VH$ $(H \to b\bar{b})$ final statistical analysis for the results published in Ref. [5]. Following the release of the analysis results in 2018, I conducted preliminary studies on the optimisation of the training parameters of the MVA analysis, documented in Sect. 7.5.1 of this thesis. These studies have served as a basis for further investigations, leading to a modification of the MVA training parameters for the following iteration of the analysis, performed with the full Run-2 dataset, corresponding to an integrated luminosity of $139 fb^{-1}$ [6]. The full Run-2 analysis is not discussed in this manuscript.

**Interpretation of the $VH$ $(H \to b\bar{b})$ cross section measurements based on an effective Lagrangian approach** The 'observation' analysis has been further extended to provide a finer interpretation of the $VH$ $(H \to b\bar{b})$ signal measurement. The cross sections for the $VH$ production times the $H \to b\bar{b}$ branching ratio have been measured in exclusive regions of phase space. I performed the interpretation of these measurements based on an effective field theory (EFT) approach. I was responsible for the whole interpretation work, from the parametrisation of the inputs to the fitting procedure and cross checks. The EFT interpretation results have been included in the corresponding ATLAS paper [7], published in 2019. I also documented the results in the internal note [8]. Chapter 8 of this thesis discusses the $VH$ $(H \to b\bar{b})$ cross section measurements, with a special focus on the EFT interpretation.

**Similarity search for the fast simulation of the Forward Calorimeter in ATLAS** In the context of a project in collaboration with the AMVA4NewPhysics ITN, I was based in Moscow for a period of 3 months. I have collaborated with researchers in Yandex and Higher School of Economics (National Research University), working on *similarity search* techniques as a new approach for the fast simulation of the forward calorimeter (FCal) in ATLAS. Such techniques constitute a branch of machine learning algorithms and include clustering and indexing methods that enable quick and efficient electromagnetic shower simulation in the FCal. I have developed and implemented the new approach in the simulation framework used by the ATLAS experiment (Athena). The new solution outperforms the standard fast simulation technique used in ATLAS. The fast simulation project is described in Chap. 6.

This research project has been supported by a Marie Skłodowska-Curie Innovative Training Network Fellowship of the European Commission's Horizon 2020 Programme under contract number 675440 AMVA4NewPhysics.

Chicago, USA Cecilia Tosciri

## References


1. AMVA4NewPhysics authors. D1.1 Multivariate Analysis Methods for Higgs Boson Searches at the Large Hadron Collider. url: https://amva4newphysics.wordpress.com/deliverables/
2. Tosciri C, Bortoletto D (2017) Classification studies for Higgs pair production in the resolved $b\bar{b}b\bar{b}$ final state. Tech Rep ATL-COM-PHYS-2017-1628. Geneva: CERN. url: https://cds.cern.ch/record/2290881
3. Tosciri C. BJetJER20p7. url: https://twiki.cern.ch/twiki/bin/viewauth/AtlasProtected/BJetJER20p7
4. Buzatu A, Jiggins S (2018) Object definitions and selections for standard model $VH \rightarrow ll/lv/vv + b\bar{b}$analysis. Tech Rep ATL-COM-PHYS-2018-517. Geneva: CERN. url: https://cds.cern.ch/record/2317182
5. ATLAS Collaboration (2018) Observation of $H \rightarrow b\bar{b}$ decays and $VH$ production with the ATLAS detector. Phys. Lett. B786:59–86. doi: https://doi.org/10.1016/j.physletb.2018.09.013. arXiv: 1808.08238 [hep-ex]
6. Aad G et al (2021) Measurements of $WH$ and $ZH$ production in the $H \rightarrow b\bar{b}$ Decay channel in pp collisions at 13 TeV with the ATLAS detector. Eur Phys J C 81(2):178. doi: https://doi.org/10.1140/epjc/s10052-020-08677-2. arXiv: 2007.02873 [hep-ex]
7. Aaboud M et al (2019) Measurement of $VH$, $H \rightarrow b\bar{b}$ production as a function of the vector-boson transverse momentum in 13 TeV pp collisions with the ATLAS detector. JHEP 05:141. doi: 10.1007/JHEP05(2019)141. arXiv: 1903.04618 [hep-ex]
8. Calvet TP et al (2018) Simplified template cross-section measurements for the V H($H \rightarrow b\bar{b}$) process with the ATLAS detector. Tech Rep ATL-COM-PHYS-2018-1205. Geneva: CERN. url: https://cds.cern.ch/record/2636121

# Acknowledgements

I am sincerely grateful to all the people that contributed to the realisation of my D.Phil. and the completion of this thesis.

I am deeply grateful to my supervisor Daniela Bortoletto for giving me the unique opportunity to work in the most exciting international environments for scientific research, as the University of Oxford and CERN are. I thank Daniela for her invaluable guidance and advice and for giving me the possibility to expand my research interests and pursue many different compelling projects.

I would like to thank all the members of the ATLAS $H \to b\bar{b}$ group and $VH(H \to b\bar{b})$ analysis team. I have been honoured to contribute to the production of our important results. I am very much thankful to Valerio Dao for the fruitful discussions, and precious advise and for being always open to help students. I also thank Adrian Buzatu, Giovanni Marchiori, and Paolo Francavilla.

It was a privilege to be part of the AMVA4NewPhysics network and collaborate with a group of brilliant senior researchers and students. I owe a special thanks to Tommaso Dorigo for the ability to successfully coordinate such a great team.

Thanks to the network, I had the possibility to participate in many training activities and unique collaborations, which brought me everywhere across Europe. In particular, my secondment at Yandex and HSE in Moscow was remarkable, both for the place and for the project on which I have worked. I am very thankful to Fedor Ratnikov for welcoming and introducing me into the project. A special thanks goes to Sasha Glazov, who guided me throughout the technicalities of the project to its final realisation.

Thanks to the Higgs group in Oxford, in particular to Cigdem Issever, Chris Hays, and Gabija Zemaytite, for their advice and help in some decisive steps of my D.Phil.

Thanks to all the participants of the Bortoletto-Shipsey meetings for being of great inspiration for me with their insightful projects. A special thanks to Philipp Windischhofer for providing useful feedback on the content of this thesis.

Thanks to the past and present fellow Oxford students, especially to Maryian Petrov and Nurfikri Norjoharuddeen, who filled our office with their colourful personalities during the first years of this D.Phil. A special thanks to Luca Ambroz, my great office mate, colleague and friend, who shared with me joys and hopes of this extraordinary experience.

A big thanks to the administration team of the Particle Physics Department for their professionalism and kindness. Thank you to Sue Geddes, Jennifer Matthews, and Kim Proudfoot.

I am deeply grateful to my family. Despite the distances, they are my greatest motivation and are always in my thoughts.

Finally, I am incredibly grateful to Marco Del Tutto for his support and encouragement, both personal and professional. You are a wonderful person and an extraordinarily talented physicist, and this thesis is dedicated to you.

# Contents

# Chapter 1
# Introduction

The most accurate description of Nature conceived by the humankind is enclosed in the formulation of the Standard Model (SM) of particle physics. The construction of the SM started in the 1950's to extend a successful yet incomplete theory, Quantum Electrodynamics (QED), and was driven by brilliant ideas, such as the quark model, the gauge symmetry and its spontaneous breaking [1–4]. The success of the SM came with its experimental validation, characterised by many particle discoveries and technological breakthroughs [5]. The observation in 2012 of the Higgs boson at the Large Hadron Collider (LHC) marks the culmination and triumph of the SM, as the last missing piece predicted by the model was finally discovered. Since then, the properties of the Higgs boson have been measured with increased precision and proved to be consistent with the predictions. Nevertheless, despite the success of the SM, it is now accepted by the scientific community that, similarly to how the QED was an approximation of the SM, also the SM must be only an effective manifestation of a more fundamental description of Nature.

To date, many physical phenomena are not explained and several questions remain unresolved, from the nature of neutrinos and dark matter to the origin of the baryon asymmetry in the Universe. The properties of the Higgs boson are also still under investigation. Is the Higgs boson a fundamental scalar or a more complex object? Why is the Higgs mass so many orders of magnitude lower than the straightforward prediction (leading to the hierarchy or naturalness problem in the SM [6])?

The physics program at the LHC is strongly committed to extend the SM. The search for hints of physics beyond the Standard Model (BSM) can either be pursued directly through searches for production of new particles or indirectly, by looking at possible divergencies from the SM. Direct searches are based on the assumptions of specific BSM theoretical models, such as the *supersymmetric theories* (SUSY) [7] or the two-Higgs-doublet model (2HDM) [8]. With an indirect approach, precise measurements of cross sections are compared to SM predictions in search of deviations.



C. Tosciri, *Higgs Boson Decays into a Pair of Bottom Quarks*, Springer Theses,
https://doi.org/10.1007/978-3-030-87938-9_1

In this context, an effective field theory (EFT) can be used for the interpretation of the LHC data.

The extensive physics program at the LHC requires increasingly advanced computational and data analysis techniques. Recently, machine learning methods have made a prominent appearance in the field of particle physics, and promise to convey many challenges encountered at the LHC. One of the main limits is certainly the simulation of the detector response. Physics measurements typically require extremely detailed and precise simulation that rely on the well understood physics governing the interaction of particles with matter. These simulations are generally very computational demanding and require a big amount of the computing resources of an experiment. Innovative machine learning techniques for fast simulations are currently under investigation, when not already in use, at the LHC experiments.

This thesis presents the results of the analysis that in 2018 led to the observation of the Higgs boson decaying into pairs of bottom quarks. The analysis exploits the production of a Higgs boson associated with a vector boson. The leptonic decays of the vector boson provide signatures that enable an efficient triggering and a powerful background reduction. The main analysis strategy is based on a multivariate approach that maximises the signal sensitivity by discriminating the signal events from the background in a powerful way. The analysis is performed on a dataset of proton-proton (*pp*) collisions at a centre-of-mass energy of 13 TeV, collected by the ATLAS experiment during Run-2 and corresponding to a luminosity of 79.8 $fb^{-1}$. The analysis results are provided in terms of signal strength, the ratio between the measured signal yield and the SM expectation.

The extension of the $VH(H \rightarrow b\bar{b})$ analysis with further interpretation of the signal measurement is also described in this manuscript. The cross sections for the $VH$ production times the $H \rightarrow b\bar{b}$ branching ratio as a function of the vector boson transverse momentum is measured in exclusive regions of phase space. These measurements are used to search for possible deviations from the SM using an EFT approach, based on anomalous couplings of the Higgs boson.

A novel technique for the fast simulation of the forward calorimeter response based on *similarity search* is also presented.

The structure of the thesis is briefly outlined in what follows.

Chapter 2 introduces the theoretical argumentations of the SM and the Higgs mechanism, fundamental for putting in context the following chapters.

Chapter 3 introduces the concepts necessary for the implementation and interpretation of the machine learning techniques used in this thesis.

Chapter 4 is an overview of the LHC and the ATLAS functioning.

Chapter 5 describes the principal algorithms used for the reconstruction of physics objects at the ATLAS experiment.

Chapter 6 presents the novel technique developed for the fast simulation of the forward calorimeter in ATLAS, based on similarity search.

Chapter 7 presents the $VH(H \rightarrow b\bar{b})$ ATLAS analysis performed with 79.8 $fb^{-1}$.

Chapter 8 presents the $VH(H \rightarrow b\bar{b})$ cross section measurements, including an effective field theory interpretation.

Chapter 9 provides a brief summary and conclusions.

## References


1. Perkins DH (1982) Introduction to high energy physics. ISBN: 9780521621960
2. Halzen F, Martin AD (1984) Quarks and leptons: an introductory course in modern particle physics. ISBN: 0471887412, 9780471887416
3. Schwartz MD (2014) Quantum field theory and the standard model. Cambridge University Press, Mar 2014. ISBN: 978-1-107-03473-0, 978-1-107-03473-0
4. Thomson M (2013) Modern particle physics. Cambridge University Press. https://doi.org/10.1017/CBO9781139525367
5. Zyla PA, others (Particle Data Group) (2020) Review of particle physics. In: PTEP 2020.8, p 083C01. https://doi.org/10.1093/ptep/ptaa104
6. 't Hooft G (1980) Naturalness, chiral symmetry, and spontaneous chiral symmetry breaking. In: 't Hooft G et al (eds), NATO science series B, vol 59, pp 135–157. https://doi.org/10.1007/978-1-4684-7571-5_9
7. Sohnius MF (1985) Introducing supersymmetry. Phys Rept 128:39–204. https://doi.org/10.1016/0370-1573(85)90023-7
8. Branco GC et al (2012) Theory and phenomenology of two-Higgs-doublet models. Phys Rept 516 :1–102. https://doi.org/10.1016/j.physrep.2012.02.002. arXiv:1106.0034 [hep-ph]

# Chapter 2
# Theoretical Introduction

The SM of particle physics [1–3] describes the fundamental constituents of the Universe and the forces that regulate their interactions. The SM has now been completed since a particle compatible with the SM Higgs boson has been recently discovered. Despite the remarkable agreement between the model and the experimental observations, it is reasonable to expect that the properties of the SM particles, such as the Higgs boson, could be perturbed by physics beyond our current understanding.

This chapter gives an overview of the SM formulation, with a special focus on the Higgs mechanism, as well as a summary of the Higgs boson properties measurable at the LHC. The EFT approach is also introduced as a framework for studying possible deviations from the SM predictions.

## 2.1 Standard Model

The SM predicts the existence of a small number of fundamental particles of spin $\frac{1}{2}$, called *fermions*, and arranged in a scheme of three generations, in such a way that the *second* and *third generations* can be seen as heavier replicas of the *first generation*, as illustrated in Fig. 2.1. Fermions are further categorised according to their quantum numbers. Fermions with a colour charge and a fractional electric charge are called *quarks*. The up-type quarks carry an electric charge of $+\frac{2}{3}|e|$ and are identified in three flavours: up ($u$), charm ($c$), top ($t$). Similarly, the down-type quarks, which carry a negative charge of $-\frac{1}{3}|e|$, stand out in three flavours: down ($d$), strange ($s$), and bottom ($b$). *Leptons* are fermions with neutral or integral electric charge ($|e|$) and are not associated to a colour charge. Each type (or *flavour*) of charged leptons, which include electron ($e$), muon ($\mu$) and tau-lepton ($\tau$), is paired to a neutral lepton, referred to as *neutrino*, and divided into $\nu_e$, $\nu_\mu$, and $\nu_\tau$.



C. Tosciri, *Higgs Boson Decays into a Pair of Bottom Quarks*, Springer Theses,
https://doi.org/10.1007/978-3-030-87938-9_2

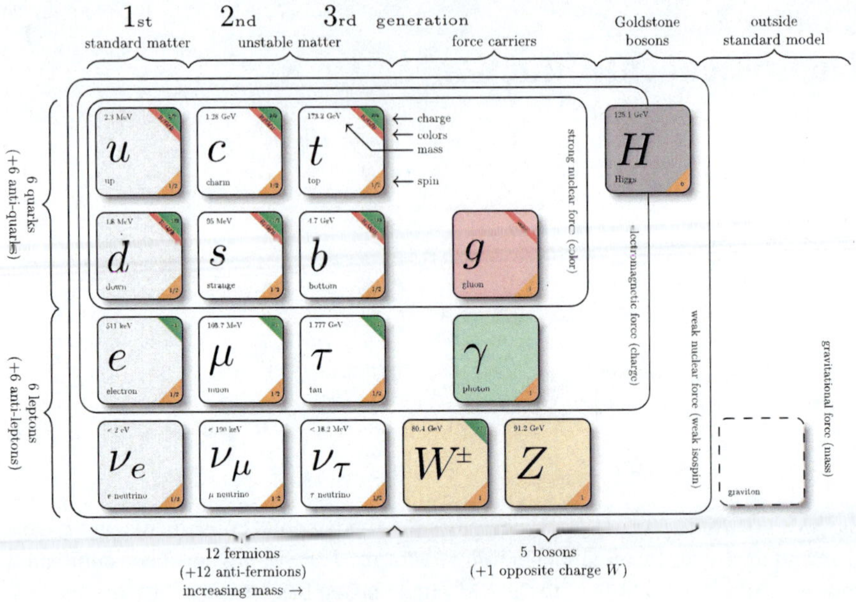


**Fig. 2.1** SM infographic. Together with the fermions, arranged in three generations, the gauge bosons, responsible for mediating the interactions, are shown. For each particle, the charge, the mass, and the spin are indicated. Image source [4]

Elementary particles interact with each other through the four fundamental forces, the *strong*, the *electromagnetic*, the *weak*, and the *gravitational* force, which explain all the experimentally observed phenomena. However, the impact of the gravitational force on particle processes is negligible, therefore this interaction is not included in the SM formulation. The interaction between fermions is mediated by force carriers of spin 1, known as *gauge bosons*. These particles are the photon $\gamma$, which mediates the electromagnetic force, the gauge bosons of the weak interaction, $W^+$, $W^-$, and $Z^0$, and finally the carrier of the strong force, the gluon $g$. Quarks are subject to the three kinds of interactions, while charged fermions interact via electromagnetic and weak interactions and neutrinos only interact weakly.

The SM is a Quantum Field Theory (QFT) [5], in which the particles are described by perturbations of quantum fields and satisfy the appropriate quantum mechanical field equations. The theory is constructed upon the Lagrangian formalism as a generalisation of Quantum Electrodynamics (QED) [6]. The dynamics of the free fermions are described by the Dirac equation of relativistic quantum mechanics. The Dirac equation has four solutions, two of which imply negative energies. These solutions are interpreted as positive energy particles propagating backwards in time or, equivalently, as positive energy *antiparticles*, objects with identical mass and lifetime as the particles introduced above, but with opposite charge, propagating forward in time. The Lagrangian density associated to the Dirac equation is:

$$\mathscr{L}_{Dirac} = \bar{\psi}(i\gamma^{\mu}\partial_{\mu} - m)\psi, \tag{2.1}$$

where $\psi = \psi(x)$ represents the field spinor, $\gamma^{\mu}$ represent the Dirac $\gamma$-matrices and $m$ is the mass of the fermion. The Lagrangian for a Dirac field is required to be invariant under the U(1) local transformation:

$$\psi(x) \rightarrow \psi'(x) = e^{iq\chi(x)}\psi(x), \tag{2.2}$$

where $q$ is the charge of the particle annihilated by the field $\psi(x)$, and $\chi(x)$ is a real differentiable function. The invariance is obtained by replacing the derivative $\partial_{\mu}$ with the covariant derivative $D_{\mu}$, defined as

$$\partial_{\mu} \rightarrow D_{\mu} = \partial_{\mu} + iqA_{\mu}, \tag{2.3}$$

where $A_{\mu}$ is a vector field, which is identified with the massless photon, invariant under the gauge transformation

$$A_{\mu} \rightarrow A'_{\mu} = A_{\mu} - \partial_{\mu}\chi. \tag{2.4}$$

This allows to define the QED Lagrangian density, which describes the fermion field, the photon field, and their interaction, in the form

$$\mathscr{L}_{QED} = \bar{\psi}(i\gamma^{\mu}\partial_{\mu} - m)\psi + e\bar{\psi}\gamma^{\mu}\psi - \frac{1}{4}F_{\mu\nu}F^{\mu\nu}, \tag{2.5}$$

where $F_{\mu\nu} = \partial_{\mu}A_{\nu} - \partial_{\nu}A_{\mu}$ is the electromagnetic field tensor.

Quantum Chromodynamics (QCD) [7] describes the strong interaction and is obtained by extending the local gauge invariance U(1) of the QED to the SU(3) local gauge symmetry. QCD is mediated by eight massless gluons, corresponding to the generators of SU(3), represented by $3 \times 3$ matrices. QCD is characterised by three *colour* charges, each colour corresponding to an orthogonal state in the SU(3) space. Quarks carry colour charge while leptons do not, and therefore only quarks can couple to gluons and interact through the strong interaction. QCD is invariant under SU(3) local phase transformations, which concerns rotations of states in the colour space. Consequently, the strength of QCD does not depend on the colour charge of the quark.

Quarks are found to be arranged in composite particles and do not exist as free particles. This phenomenon is known as *colour confinement* and regards the objects with non-zero colour charge, which can not propagate as free particles. The composite particles formed by quarks are called *hadrons*, which are categorised into *baryons*, made of three quarks, and *mesons*, made of one quark and one antiquark. The mechanism by which quarks and gluons produced in hard processes form the hadrons is called *hadronisation*. Another important property of QCD is *asymptotic freedom*, which is related to the significant variation of the coupling constant $\alpha_s$ of the strong interaction, over the range of energies relevant to particle physics. In particular, $\alpha_s$ has large values for small energy scales (~1 GeV), which correspond to large distances, making the theory non-perturbative. For higher energies, at the typical scale of collider physics (>100 GeV) the coupling strength decreases, allowing perturbative methods to be used for quantitative predictions. However, even in this condition, the convergence of the perturbative expansion is very slow and requires higher-order corrections. A specific class of corrections, the quark self-energy corrections, could lead to infinities, known as *ultraviolet divergencies*, which can be eliminated by introducing a redefinition of fields or parameters to absorb the divergent term. This procedure is called *renormalisation* and requires the introduction of a *renormalisation scale* $\mu_R$ and a *factorisation scale* $\mu_F$ [8–10]. For all these reasons, QCD calculations for processes at the LHC are very challenging.

The SM unifies the weak and electromagnetic forces into a single electroweak (EW) interaction. The weak charged-current interaction is associated with an $SU(2)_L$ gauge symmetry, whose generators are the weak isospin $\boldsymbol{T}$ that can be written in terms of Pauli matrices $\boldsymbol{\sigma}$ as $\boldsymbol{T} = \frac{1}{2}\boldsymbol{\sigma}$. The subscript $L$ of the $SU(2)_L$ group indicates that the weak charged-current interaction only couples with left-handed (LH) particle states (and right-handed (RH) antiparticle states). In this description, the LH particles and RH antiparticles are represented by *weak isospin doublets*, with total weak isospin $I_W = \frac{1}{2}$, while the RH particle and LH antiparticle states are placed in *weak isospin singlets* with $I_W = 0$. The local gauge invariance can be satisfied by introducing three gauge fields, $W^k_\mu$, with $k = 1, 2, 3$, corresponding to three gauge bosons. Linear combinations of these fields give rise to weak charged-currents, corresponding to the physical $W$ bosons:

$$W^{\pm}_\mu = \frac{1}{\sqrt{2}}(W^{(1)}_\mu \mp i W^{(2)}_\mu). \tag{2.6}$$

The $SU(2)_L$ gauge symmetry also implies the existence of a weak neutral-current that can only couple with LH particles and therefore it can not be identified with the $Z$ boson, which couples with both LH and RH particle states.

An additional symmetry $U(1)_Y$ introduces the *weak hypercharge* $Y = 2(Q - I_W)$, where $Q$ is the electromagnetic charge. A new gauge field $B_\mu$ couples to $Y$. In this model, called GSW after Glashow [11], Salam [12] and Weinberg [13], the physical photon and $Z$ boson result from linear combinations of the fields $B_\mu$ and $W^{(3)}_\mu$:

$$\begin{aligned} A_\mu &= +B_\mu \cos\theta_W + W^{(3)}_\mu \sin\theta_W, \\ Z_\mu &= -B_\mu \sin\theta_W + W^{(3)}_\mu \cos\theta_W, \end{aligned} \tag{2.7}$$

where $\theta_W$ is the weak mixing angle. Within the GSW model, the couplings of the weak and electromagnetic interactions are related as follows

$$e = g_W \sin\theta_W = g_Z \sin\theta_W \cos\theta_W. \tag{2.8}$$

Finally, the SM description unifies the EW and QCD theories under the symmetry group $SU(3)_C \otimes SU(2)_L \otimes U(1)_Y$. The required local gauge symmetry can only be satisfied if all the particles are massless, as the introduction of mass terms in the SM Lagrangian would break the local gauge invariance. This is in contrast with the experimental evidence of massive particles. It is then necessary to include in the EW sector of the SM a mechanism that spontaneously breaks the symmetry and gives rise to the mass of the particles: the *Brout–Englert–Higgs mechanism* [14–17].

## 2.2 Higgs Mechanism

According to the Higgs mechanism, the gauge bosons acquire mass upon their interaction with the Higgs field. The new field is identified with a weak isospin doublet of complex scalar fields with four degrees of freedom:

$$\phi = \begin{pmatrix} \phi^+ \\ \phi^0 \end{pmatrix} = \frac{1}{\sqrt{2}} \begin{pmatrix} \phi_1 + i\phi_2 \\ \phi_3 + i\phi_4 \end{pmatrix}, \tag{2.9}$$

where $\phi^0$ and $\phi^+$ are a neutral field and a positively charged field, respectively. The corresponding Lagrangian is

$$\mathscr{L} = (D_\mu \phi)^\dagger (D^\mu \phi) - V(\phi), \tag{2.10}$$

where the derivative $\partial_\mu$ is substituted by the covariant derivative that ensures invariance under the $SU(2)_L \otimes U(1)_Y$ local gauge symmetry:

$$\partial_\mu \rightarrow D_\mu = \partial_\mu + i g_W \boldsymbol{T} \cdot \mathbf{W}_\mu + i g' \frac{Y}{2} B_\mu. \tag{2.11}$$

The first term of the Lagrangian in Eq. (2.10) refers to the kinetic energy, while $V(\phi)$ is the Higgs potential, defined as

$$V(\phi) = \mu^2 \phi^\dagger \phi + \lambda (\phi^\dagger \phi)^2. \tag{2.12}$$

The first term of $V(\phi)$ represents the mass term, while the second introduces a self-interaction of the field. The vacuum state corresponds to the minimum of the potential. In order to obtain a finite minimum, the coefficient $\lambda$ must be positive, while $\mu^2$ can be either positive or negative. If $\mu^2 > 0$ the resulting potential has a unique minimum at $\phi = 0$, as illustrated in Fig. 2.2a. In this case, the Lagrangian in Eq. (2.10) represents a scalar particle with mass $\mu$ under the SU(2) symmetry. On the other hand, if $\mu^2 < 0$ the potential has a set of degenerate minima on a circumference corresponding to

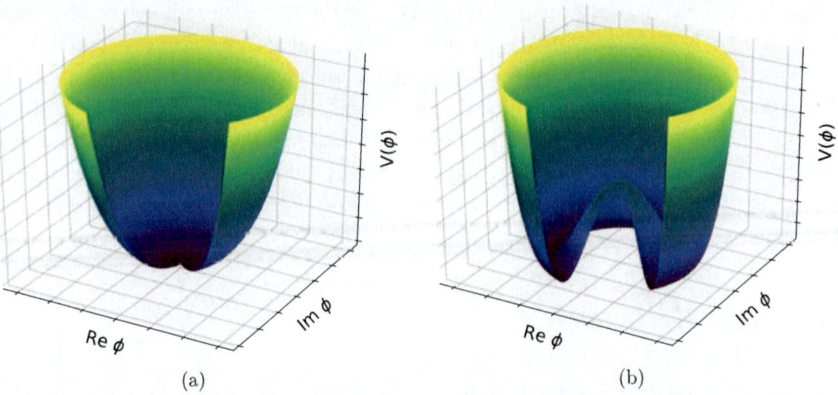


**Fig. 2.2** $V(\phi) = \mu^2\phi^\dagger\phi + \lambda(\phi^\dagger\phi)^2$ potential for a complex scalar field, in case $\mu^2 > 0$ (**a**) and $\mu^2 < 0$ (**b**). The former case has a unique minima in $\phi = 0$, while the latter corresponds to the Higgs potential, and presents a set of degenerated minima

$$\phi^\dagger\phi = \frac{1}{2}(\phi_1^2 + \phi_2^2 + \phi_3^2 + \phi_4^2) = \frac{v^2}{2} = -\frac{\mu^2}{2\lambda}. \tag{2.13}$$

The latter case is shown in Fig. 2.2b. All the minima are connected by an SU(2) transformation and are equivalent. The choice of the vacuum state breaks the symmetry of the Lagrangian: the process is known as *spontaneous symmetry breaking*. To break the symmetry, one can choose $\phi_1 = \phi_2 = \phi_4 = 0$ and only the neutral scalar field $\phi_0$ corresponds to a vacuum expectation different from zero:

$$\langle 0|\,\phi\,|0\rangle = \frac{1}{\sqrt{2}}\begin{pmatrix}0\\ v\end{pmatrix}. \tag{2.14}$$

An excitation of the fields around the vacuum expectation can be written as:

$$\phi(x) = \frac{1}{\sqrt{2}}\begin{pmatrix}\phi_1(x) + i\phi_2(x)\\ v + i\eta(x) + i\phi_4(x)\end{pmatrix}, \tag{2.15}$$

where the field $\phi$ is parametrised in terms of a real physical field $\eta(x)$ and three unphysical fields $\phi_{1,2,4}$, that correspond to the Goldstone fields. When a particular gauge transformation called *unitary gauge* is chosen, the Goldstone fields are absorbed by the vector fields and the field $\phi$ can be written as a function of the massive scalar Higgs field $h(x)$:

$$\phi(x) = \frac{1}{\sqrt{2}}\begin{pmatrix}0\\ v + h(x)\end{pmatrix}. \tag{2.16}$$

In the unitary gauge, the Lagrangian in Eq. (2.10) is expressed as:

$$
\begin{aligned}
\mathscr{L}_H = & \frac{1}{2}(\partial_\mu h)(\partial^\mu h) + \frac{1}{4} g_W^2 (W_\mu^+ W^{\mu -})(v+h)^2 \\
& + \frac{g_W^2 + g'^2}{8} (Z_\mu Z^\mu)(v+h)^2 \\
& + \frac{\mu^2}{2}(v+h)^2 - \frac{\lambda}{16}(v+h)^4
\end{aligned}
\tag{2.17}
$$

where the massive neutral gauge boson, identified with the $Z$ boson, is expressed by a linear combination of massless bosons:

$$
Z_\mu = \frac{g_W W_\mu^{(3)} - g' B_\mu}{\sqrt{g_W^2 + g'^2}}. \tag{2.18}
$$

The neutral gauge boson $A_\mu$ can be written in a similar way as a linear combination of $W_\mu$ and $B_\mu$:

$$
A_\mu = \frac{g' W_\mu^{(3)} + g_W B_\mu}{\sqrt{g_W^2 + g'^2}}. \tag{2.19}
$$

In Eq. (2.17), the terms that are quadratic with respect to the gauge fields $W^\pm$, $Z$ and the Higgs field $H$, correspond to the mass terms of the corresponding bosons:

$$
m_W = \frac{g_W v}{2}, \quad m_Z = \frac{v}{2}\sqrt{g_W^2 + g'^2} = \frac{m_W}{\cos\theta}, \quad m_H = \sqrt{2\mu^2}. \tag{2.20}
$$

The gauge boson $A_\mu$, which is identified with the photon, is not associated to any mass term and the corresponding particle is therefore massless.

The Higgs field also interacts with fermions via a Yukawa interaction, providing them with mass. In the case of leptons, and taking the electron as an example, such interaction can be written as:

$$
\mathscr{L}_{e\phi} = Y_e \left( \bar{l}_F \phi e_R + \bar{e}_R \phi^\dagger l_F \right) \tag{2.21}
$$

where $l_F$ denotes the left-handed lepton doublet, $e_R$ the right-handed singlet electron, $\phi$ is the Higgs field and $Y_e$ represents the Yukawa coupling constant between the electron and the Higgs boson. The Yukawa Lagrangian satisfies the $SU(2)_L \otimes U(1)_Y$ gauge symmetry of the SM. After spontaneous symmetry breaking, the Eq. (2.21) becomes:

$$
\mathscr{L}_{e\phi} = \frac{Y_e v}{\sqrt{2}} (\bar{e}_L e_R + \bar{e}_R e_L) - \frac{Y_e h}{\sqrt{2}} (\bar{e}_L e_R + \bar{e}_R e_L) \tag{2.22}
$$

where $e_L$ and $e_R$ refer to the left- and right-handed state of the electron, defined as:

$$e_L = \frac{1-\gamma^5}{2}\psi_e \qquad e_R = \frac{1+\gamma^5}{2}\psi_e. \tag{2.23}$$

By defining the coupling parameter as

$$Y_e = \sqrt{2}\frac{m_e}{v} \tag{2.24}$$

where $m_e$ is the mass of the electron, the Lagrangian in Eq. (2.21) takes the form:

$$\mathcal{L}_{e\phi} = -m_e\bar{e}e - \frac{m_e}{v}\bar{e}eh \tag{2.25}$$

where $\bar{e}e = \bar{e}_L e_R + \bar{e}_R e_L$. The first term contains the mass of the electron, and the second term represents the coupling of the electron with the Higgs field. While this shows how the mass term arises for the electron, the same mechanism applies to the second and third generation leptons.

In the case of quarks, the mass term can be constructed in a similar way, but with the addition of a second term:

$$\mathcal{L}_{q\phi} = -Y^d_{ij}\bar{Q}_{Li}\phi d_{Rj} - Y^u_{ij}\bar{Q}_{Li}\epsilon\phi^* u_{Rj} + \text{h.c.} \tag{2.26}$$

where $Q_{Li}$ is the left-handed quark doublet for generation $i$, and $d_{Rj}$ and $u_{Rj}$ are the right-handed down- and up-type quark singlets for generation $j$ in the weak-eigenstate basis. $Y^{u,d}_{ij}$ are $3 \times 3$ complex matrices, and $\epsilon$ is the $2 \times 2$ antisymmetric tensor. A sum over $i$ and $j$ is intended for $i, j = 1, 2, 3$.

While the first term in Eq. (2.26) is basically the same as the one in Eq. (2.21), in Eq. (2.26) we also have a second term, which is the only other invariant term one can construct under the $SU(2)_L \otimes U(1)_Y$ symmetry. This term was not present for the leptons because it would assume the presence of a $\nu_R$ field.

When $\phi$ acquires a vacuum expectation value as in Eq. (2.16), the Lagrangian for the quark masses becomes

$$\mathcal{L}_{q,\text{mass}} = -\bar{d}_L\frac{Y^d v}{\sqrt{2}}d_R - \bar{u}_L\frac{Y^u v}{\sqrt{2}}u_R + \text{h.c.} \tag{2.27}$$

Here, the matrices $Y^d$ and $Y^u$ are written in the weak-eigenstate basis, and they are not necessarily diagonal. This introduces a difference between the weak-eigenstate basis and the mass-eigenstate basis, where those matrices are diagonal. The physical states are obtained by diagonalising $Y^{u,d}$ by four unitary matrices $V^{u,d}_{L,R}$ as $M^f_{\text{diag}} = V^f_L Y^f V^{f\dagger}_R (v/\sqrt{2})$ [18]. After applying these four unitary matrices, it is found that the up-type quarks diagonalise the mass, while a misalignment remains for the down-type quarks, and it is possible to go from the mass-eigenstate basis (also referred to as physical basis) to the weak-eigenstate basis via the Cabibbo–Kobayashi–Maskawa (CKM) matrix defined as $V_{\text{CKM}} = V^u_L V^{d\dagger}_L$:

$$d_L = V_{\text{CKM}}(d_{\text{ph}})_L \tag{2.28}$$

where $(d_{\text{ph}})_L$ is the down-type quark in the physical basis. Using the physical basis, the Lagrangian for the quark masses is now diagonal:

$$\begin{aligned}\mathscr{L}_{q,\text{mass}} &= -(\bar{d}_{\text{ph}})_L m_d (d_{\text{ph}})_R - (\bar{u}_{\text{ph}})_L m_u (u_{\text{ph}})_R + \text{h.c.} \\ &= -\bar{d}_{\text{ph}} m_d d_{\text{ph}} - \bar{u}_{\text{ph}} m_u u_{\text{ph}}.\end{aligned} \tag{2.29}$$

The implication of the CKM matrix is that it is possible to have a coupling between $u_{Lj}$ and $d_{Lk}$ quarks, where $j$ and $k$ are two different generations, through the charged-current $W^{\pm}$ interactions.

The masses of the particles resulting from the construction of the model are not known a priori, and are considered as free input parameters of the SM.

## 2.3 Cross Section and Decay Rate

The description of the particle production phenomenology at the LHC [19] relies on the definition of cross section in hadronic collisions. The treatment of the hadronic cross section is based on the QCD *parton model*. Within a hadron scattering, the hard collision takes place between two partons $a$ and $b$ (quarks or gluons) which carry momentum fractions $x_1$ and $x_2$ of the incoming hadrons. The partonic cross section $\hat{\sigma}_{ab\to X}$ describes the probability of a transition from the initial state of the partons to a given final state $X$ and is computed from the interaction Lagrangian. According to the *factorisation theorem* [10], the inclusive cross section for the specific process $pp \to X$ is given by:

$$\sigma_{pp\to X} = \sum_{a,b} \int_0^1 dx_1 \int_0^1 dx_2 f_a(x_1, \mu_{\text{F}}) f_b(x_2, \mu_{\text{F}}) \int d\hat{\sigma}_{ab\to X}(x_1 P_1, x_2 P_2, \mu_{\text{R}}, \mu_{\text{F}}), \tag{2.30}$$

where the *parton distribution functions* (PDFs) $f_i(x_i, \mu_{\text{R}}, \mu_{\text{F}})$ represent the probability for a parton to participate in the hard scattering interaction with momentum $p_i = x_i P_i$, within a proton of momentum $P_i$. The PDFs are independent from the hard process and are characterised by small energy scales. The separation of the soft and hard domains is performed at the *factorisation* scale $\mu_F$, which represents the maximum value of the transverse momentum carried by hadronic particles that are considered as part of the hadron [10].

The process $ab \to X$ can refer to a specific Higgs production process. The computation of the inclusive production cross sections for the Higgs boson $pp \to H$ is essential for the proper interpretation of the experimental measurements. However, at the LHC only measurements of partial cross sections for specific Higgs boson production and decay channels $pp \to H \to \text{Y}$ can be performed. The *branching ratio* $BR(H \to Y) = \Gamma_{H \to Y}/\Gamma_H$ represents the probability for the Higgs boson to decay into the final state Y. The BR of each decay mode is defined in relation to the total decay width of the Higgs boson $\Gamma_H$. Precise predictions for the specific decay widths are needed to properly weight the Higgs boson production cross section for the exclusive channel $H \to Y$.

## 2.4 Properties of the Higgs Boson

According to the SM, the Higgs boson is a neutral scalar particle which couples to fermions and gauge bosons with a coupling strength proportional to the mass of the particles. As a consequence, the Higgs boson is preferably produced in processes that involve heavy particles and, similarly, it preferably decays into the more massive particles that are kinematically accessible.

The main production mechanisms for Higgs bosons [20] in *pp* collisions at the LHC are illustrated by Leading Order (LO) diagrams in Fig. 2.3. The dominant production mode is the gluon fusion *ggF*. Since the Higgs boson does not couple with massless particles such as the gluons, the *ggF* production proceeds through a loop diagram involving a virtual top quark. Loops with lighter particles are suppressed. Although the *ggF* process has the largest cross section, from an experimental point

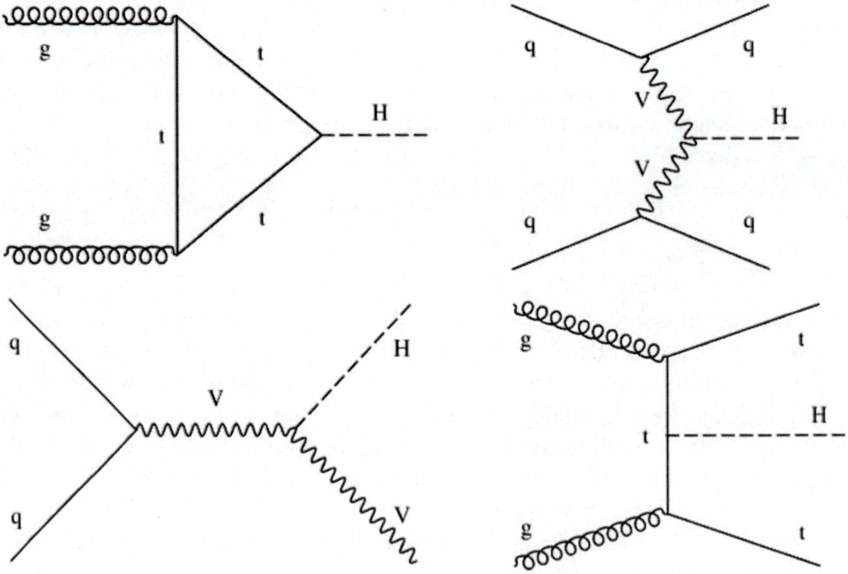


**Fig. 2.3** Examples of LO Feynman diagrams for the main Higgs boson production modes: gluon fusion (top left), vector boson fusion (top right), associated production with a vector boson *V* (bottom left) and with a pair of top quarks (bottom right). Image source [21]

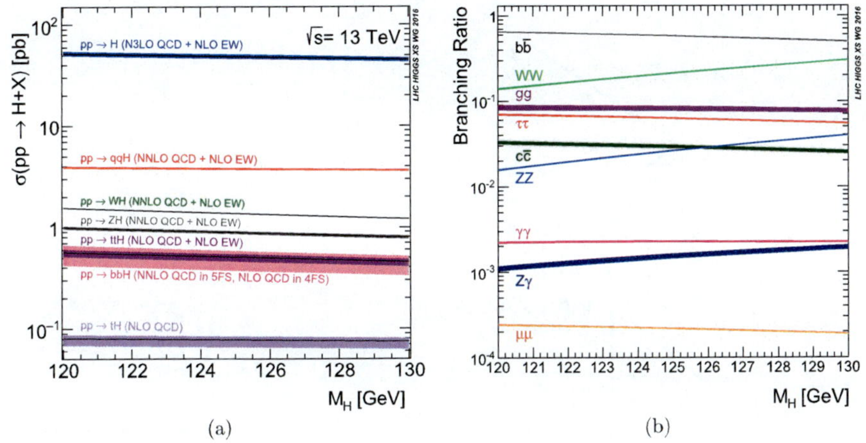


**Fig. 2.4** Cross sections of the Higgs production modes (**a**) and branching ratios of the main decay modes (**b**) in *pp* colliders as a function of the Higgs boson mass value. Image source [22]

of view it is not the easiest channel for studying the Higgs boson, since the final state is largely contaminated by QCD radiation.

The production mode with the second-largest cross section is called *vector boson fusion* (VBF) and involves the scattering of two quarks, each mediated by a vector boson $V$ ($W^{\pm}$ or $Z$) that produce the Higgs boson. The VBF production provides a more clean signature than $gg$F, since it is characterized by two forward jets in addition to the decay products of the Higgs boson.

Another relevant mechanism is the Higgs boson production in association with a vector boson $V$ ($VH$ production or *Higgs-strahlung*). This channel offers a very clean signature for the Higgs boson, as discussed in Chap. 7.

Another important channel is the Higgs production in association with a pair of top quarks, which offers the unique opportunity to extract information on the direct coupling of the Higgs boson to the top quark.

The cross sections of the mentioned Higgs production modes are shown in Fig. 2.4a as a function of the Higgs boson mass.

The Higgs boson is unstable and, in principle, it can decay into all the particles predicted by the SM. The decay branching ratio (BR) is determined by the mass of the involved particles. In Fig. 2.4b the BR of the decay modes accessible at the LHC are shown as a function of the Higgs mass. The dominant decay mode is $H \to b\bar{b}$, which will be discussed in Chap. 7. Examples of LO Feynman diagrams for the Higgs boson decays are shown in Fig. 2.5. The experimental observability of the various decays is affected by diverse background processes.

In 2012 a particle consistent with the scalar boson resulting from the SM Higgs mechanism was discovered at the LHC by the ATLAS [23] and CMS [24] experiments, representing a milestone for the LHC and a remarkable success of the SM. Since then, substantial progress has been made, yielding an increasingly precise infor-

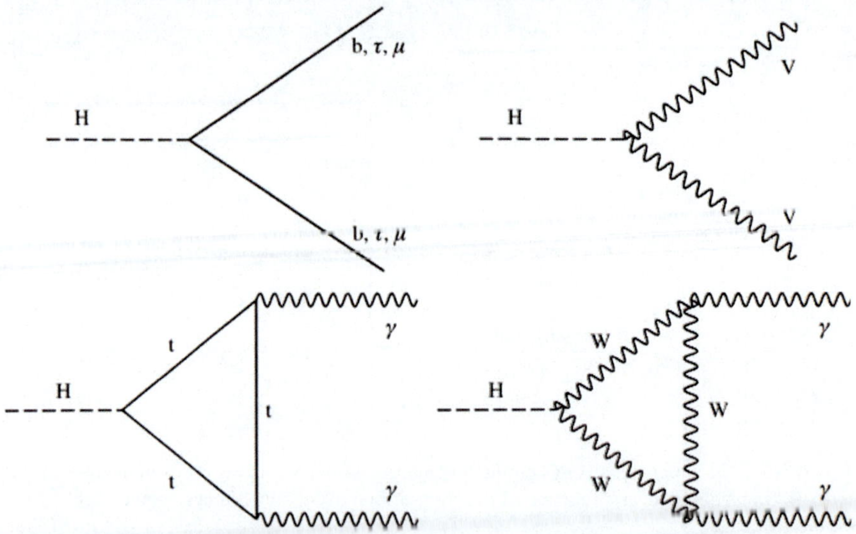


**Fig. 2.5** Examples of LO Feynman diagrams for the Higgs boson decay modes: $H \to b\bar{b}$, $H \to \tau\tau$, $H \to \mu\mu$ (top left), $H \to ZZ$, $H \to WW$ (top right), $H \to \gamma\gamma$ channels (bottom row). Image source [21]

mation on the properties of the Higgs boson. Its mass has been precisely measured ($125.18 \pm 0.16$ GeV [18]) and the accessible production and decay rates are found to be consistent with the SM predictions. The compatibility of the experimental results with the SM is expressed in Fig. 2.6 by the measured coupling-strength modifiers defined as $\sqrt{\kappa_V}\frac{m_V}{v}$ for weak bosons with a mass $m_V$, and as $\kappa_F\frac{m_F}{v}$ for fermions with mass $m_F$. All measured coupling scale factors are found to be compatible with their SM expectation (dotted line) [25].

## 2.5 Effective Field Theory

Every physics theory can be considered as an approximation of the physics at higher scales of energy, characterised by parameters that result to be small if compared to the physical quantities observed. These parameters enter the approximate description only as small perturbations [26]. A familiar example is QED, the first QFT developed in particle physics, which, in its own, represents an approximation of the SM as it is currently conceived. QED can be obtained from the SM by integrating out all the particles other than photons and electrons [27]. From this perspective, it seems reasonable to assume that the SM is an approximation of a more complex underlying physical theory. It is possible to describe the physics that we are able to test at the LHC in terms of a QFT with a finite set of parameters, without any reference to the energy scale at which the unknown underlying theory is valid. The only implication

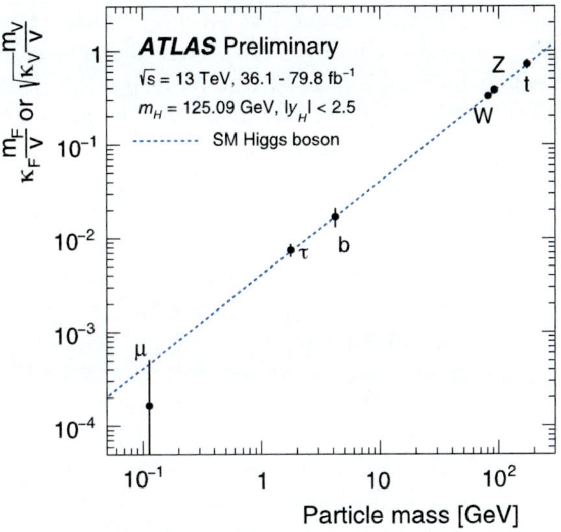


**Fig. 2.6** Measured coupling-strength modifiers defined as $\sqrt{\kappa_V}\frac{m_V}{v}$ for weak bosons and as $\kappa_F\frac{m_F}{v}$ for fermions. All the coupling scale factors are compatible with their SM expectation (dotted line). Image source [25]

is that the accessible energy scale is orders of magnitude smaller than the scale of BSM, E $\ll \Lambda$. This approach is totally model-independent, as it does not assume any specific BSM.

In the Standard Model Effective Field Theory (SMEFT) the SM Lagrangian of dimension $D = 4$ is extended by a set of local operators written in terms of SM fields, in such a way that each consecutive term is suppressed by a larger power of the energy scale of new physics $\Lambda$, which is unknown:

$$\mathscr{L}_{SMEFT} = \mathscr{L}_{SM} + \sum_i \frac{c_i^{(5)}}{\Lambda}\mathcal{O}_i^{(5)} + \sum_i \frac{c_i^{(6)}}{\Lambda^2}\mathcal{O}_i^{(6)} + \sum_i \frac{c_i^{(7)}}{\Lambda^3}\mathcal{O}_i^{(7)} + \sum_i \frac{c_i^{(8)}}{\Lambda^4}\mathcal{O}_i^{(8)} + \cdots . \tag{2.31}$$

In Eq. (2.31), $c_i$ indicate coupling constants, referred to as *Wilson coefficients*, and $\mathcal{O}_i^{(D)}$ are operators of dimension $D$, invariant under the $SU(3)_C \otimes SU(2)_L \otimes U(1)_Y$ symmetry. The operators with dimension $D = 5$ violate the lepton number [28], which is assumed to be conserved as a consequence of the SM symmetry, while the $D = 7$ operators violate baryon number minus lepton number $B - L$ [29], which should be conserved for the same reason. Therefore, the leading BSM effects come from dimension $D = 6$ operators $\mathcal{O}_i^{(6)}$, which usually describe small deviations from the SM. At this order, the model counts 59 independent operators involved in the effective Lagrangian, of which only a restricted set affects Higgs physics, and capture all the possible deviations from the SM [30].

The truncation of the SMEFT Lagrangian at dimension $D = 6$ to approximate the effective theory is accepted by the LHC community and is the main choice of many model implementations currently available. However, it is important to consider that, in some cases, non-trivial corrections can appear at least at dimension $D = 8$. For instance, the contribution of dimension $D = 8$ can be dominant when the operators mediate an interaction that does not appear at lower level in the EFT expansion [31]. A pertinent example for the work presented in this thesis is the $gg \rightarrow ZH$ production mode, which appears at Next-to-Next-to-Leading-Order (NNLO) through heavy-quark loops. This process is not affected by perturbations due to dimension $D = 6$ operators evaluated at LO, but might be subject to modifications from dimension $D = 8$ operators.

The EFT operators can be defined in several different bases, with a preference for those that are complete (meaning that all the $D = 6$ operators are included in the basis or they represent a linear combination of operators of the basis) and non-redundant (meaning that the basis is a minimal set of operators). The *Strongly-interacting light Higgs* (SILH) [32] and the *Warsaw* bases [33] are currently considered as the most convenient bases containing operators relevant for Higgs physics. The SILH basis is derived from an effective theory of a light composite Higgs boson associated with strong dynamics and responsible for EW symmetry breaking. The Warsaw basis, which is a completely renormalised theory is currently recommended over the SILH basis, which is considered incomplete [34, 35]. However, the EFT results that will be presented in Sect. 8.4 are obtained with the *Higgs Effective Lagrangian* (HEL) implementation [36] which is a partial implementation of the SILH and contains operators up to dimension $D = 6$. The reason for this choice is due to the larger availability of public implementation and tools needed when the work presented in this thesis was performed.

The parameters arising from the EFT construction can be constrained through precision measurements at the LHC. The main advantage of the EFT approach is that it is independent from specific theoretical models. This means that all the constraints set on precise measurements can be re-interpreted later as constraints on masses or couplings of new particles expected from any BSM. Global studies at Next-to-Leading-Order (NLO), which involve combinations of precision Higgs and EW measurements, are key to BSM interpretation. Nonetheless, LO constraints on EFT parameters relevant for specific processes, performed with Run-2 data, represent the first necessary step toward a comprehensive treatment of the LHC data with the EFT approach.

## References

1. Perkins DH (1982) Introduction to high energy physics. ISBN: 9780521621960
2. Halzen F, Martin AD (1984) Quarks and leptons: an introductory course in modern particle physics. ISBN: 0471887412, 9780471887416
3. Thomson M (2013) Modern particle physics. Cambridge University Press, Cambridge. https://doi.org/10.1017/CBO9781139525367
4. Galbraith D, Burgard C (2012) Go on a particle quest at the first CERN webfest. In: BUL-NA-2012-269. 35/2012 (2012), p 10. https://cds.cern.ch/record/1473657
5. Mandl F, Shaw G (1985) Quantum field theory
6. Feynman RP (1986) QED: the strange theory of light and matter. ISBN: 978-0-691- 02417-2
7. Greiner W, Schramm S, Stein E (2002) Quantum chromodynamics

8. Altarelli G (2013) Collider physics within the standard model: a primer. arXiv: 1303.2842 [hep-ph]
9. Mangano ML (1999) Introduction to QCD. In: CERN-OPEN-2000-255. https://doi.org/10.5170/CERN-1999-004.53. https://cds.cern.ch/record/454171
10. Collins JC, Soper DE, Sterman GF (1989) Factorization of hard processes in QCD. Adv Ser Direct High Energy Phys 5:1–91. https://doi.org/10.1142/9789814503266_0001. arXiv: hep-ph/0409313 [hep-ph]
11. Glashow SL (1961) Partial symmetries of weak interactions. Nucl Phys 22:579–588. https://doi.org/10.1016/0029-5582(61)90469-2
12. Salam Abdus (1968) Weak and electromagnetic interactions. Conf Proc C680519:367–377. https://doi.org/10.1142/9789812795915_0034
13. Weinberg Steven (1967) A model of leptons. Phys Rev Lett 19:1264–1266. https://doi.org/10.1103/PhysRevLett.19.1264
14. Englert F, Brout R (1964) Broken symmetry and the mass of gauge vector mesons. Phys Rev Lett 13:321–323 [157 (1964)]. https://doi.org/10.1103/PhysRevLett.13.321
15. Higgs Peter W (1964) Broken symmetries, massless particles and gauge fields. Phys Lett 12:132–133. https://doi.org/10.1016/0031-9163(64)91136-9
16. Higgs PW (1964) Broken symmetries and the masses of gauge bosons. Phys Rev Lett 13:508–509 [160 (1964)]. https://doi.org/10.1103/PhysRevLett.13.508
17. Higgs Peter W (1966) Spontaneous symmetry breakdown without massless bosons. Phys Rev 145:1156–1163. https://doi.org/10.1103/PhysRev.145.1156
18. Zyla PA, others (Particle Data Group) (2020) Review of particle physics. PTEP 2020(8):083C01. https://doi.org/10.1093/ptep/ptaa104
19. Dittmaier S, Schumacher M (2013) The higgs boson in the standard model - from LEP to LHC: expectations, searches, and discovery of a candidate. Prog Part Nucl Phys 70:1–54. https://doi.org/10.1016/j.ppnp.2013.02.001. arXiv:1211.4828 [hep-ph]
20. Murray William, Sharma Vivek (2015) Properties of the higgs boson discovered at the large hadron collider. Ann Rev Nucl Part Sci 65:515–554. https://doi.org/10.1146/annurev-nucl-102313-025603
21. Sirunyan AM, et al (2019) Combined measurements of Higgs boson couplings in proton- proton collisions at $\sqrt{s} = 13$ TeV. Eur Phys J C 79(5): 421. https://doi.org/10.1140/epjc/s10052-019-6909-y. arXiv: 1809.10733 [hep-ex]
22. de Florian D, et al (2016) Handbook of LHC higgs cross sections: 4. Deciphering the nature of the higgs sector. https://doi.org/10.23731/CYRM-2017-002. arXiv:1610.07922 [hep-ph]
23. Aad G, et al (2012) Observation of a new particle in the search for the standard model higgs boson with the ATLAS detector at the LHC. Phys Lett B 716:1–29. https://doi.org/10.1016/j.physletb.2012.08.020. arXiv: 1207.7214 [hep-ex]
24. Chatrchyan S, et al (2012) Observation of a New Boson at a Mass of 125 GeV with the CMS experiment at the LHC. Phys Lett B 716:30–61. https://doi.org/10.1016/j.physletb.2012.08.021. arXiv: 1207.7235 [hep-ex]
25. Aad G, et al (2020) Combined measurements of Higgs boson production and decay using up to 80 $\text{fb}^{-1}$ of proton-proton collision data at $\sqrt{s} = 13$ TeV collected with the ATLAS experiment. Phys Rev D 101(1):012002. https://doi.org/10.1103/PhysRevD.101.012002. arXiv: 1909.02845 [hep-ex]
26. Georgi H (1993) Effective field theory. Ann Rev Nucl Part Sci 43:209–252. https://doi.org/10.1146/annurev.ns.43.120193.001233
27. Manohar AV (2018) Introduction to effective field theories. In: Les Houches summer school: EFT in particle physics and cosmology Les Houches, Chamonix Valley, France, July 3–28, 2017. arXiv:1804.05863 [hep-ph]
28. Weinberg Steven (1979) Baryon and Lepton Nonconserving Processes. Phys Rev Lett 43:1566–1570. https://doi.org/10.1103/PhysRevLett.43.1566
29. de Gouvea A, Herrero-Garcia J, Kobach A (2014) Neutrino masses, grand unification, and baryon number violation. Phys Rev D 90:016011. https://doi.org/10.1103/PhysRevD.90.016011.https://link.aps.org/doi/10.1103/PhysRevD.90.016011

30. Falkowski A, et al (2015) Rosetta: an operator basis translator for standard model effective field theory. Eur Phys J C 75(12):583. https://doi.org/10.1140/epjc/s10052-015-3806-x. arXiv:1508.05895 [hep-ph]
31. Contino R, et al (2016) On the validity of the effective field theory approach to SM precision tests. JHEP 07:144. https://doi.org/10.1007/JHEP07(2016)144. arXiv:1604.06444 [hep-ph]
32. Giudice GF, et al (2007) The strongly-interacting light higgs. JHEP 06:045. https://doi.org/10.1088/1126-6708/2007/06/045. arXiv:hep-ph/0703164 [hep-ph]
33. Grzadkowski B, et al (2010) Dimension-six terms in the standard model Lagrangian. JHEP 10:085. https://doi.org/10.1007/JHEP10(2010)085. arXiv:1008.4884 [hep-ph]
34. Alonso R, et al (2014) Renormalization group evolution of the standard model dimension six operators III: Gauge coupling dependence and phenomenology. JHEP 04:159. https://doi.org/10.1007/JHEP04(2014)159. arXiv:1312.2014 [hep-ph]
35. Elias-Miro J, et al (2013) Higgs windows to new physics through d=6 operators: constraints and one-loop anomalous dimensions. JHEP 11:066. https://doi.org/10.1007/JHEP11(2013)066. arXiv:1308.1879 [hep-ph]
36. Alloul A, Fuks B, Sanz V (2014) Phenomenology of the higgs effective Lagrangian via FEYNRULES. JHEP 04:110. https://doi.org/10.1007/JHEP04(2014)110. arXiv: 1310.5150 [hep-ph]

# Chapter 3
# Machine Learning in High Energy Physics

Machine learning (ML) [1] refers to all algorithms and models that provide a system with the ability to learn and improve, based on the inference of statistical dependencies, without being explicitly programmed to perform a specific task. ML is perceived as one of the main disruptive technologies of our ages. In parallel to the rise in industrial applications, scientists have become increasingly interested in the potential of ML for fundamental research. For decades, ML applications have played an important role in the field of High Energy Physics (HEP) for the improvement of the sensitivity of physics analyses and reconstruction techniques [2–4]. The advent of *deep learning* [5], a modern discipline that comprises ML methods based on multi-layer artificial neural networks, together with powerful computational resources and tools become available, has triggered even more interest on ML in the HEP community.

Providing a comprehensive description of the ML techniques used by researchers at the LHC is beyond the scope of this work. However, in the following sections an overview of the main concepts used in this thesis is provided, to contextualise the different applications. In particular, the Boosted Decision Tree algorithm, described in Sect. 3.2.1, is key to enhancing the sensitivity of the main analysis presented in Chap. 7. Moreover, similarity searches, introduced in the context of clustering algorithms in Sect. 3.4.2, are at the core of the fast simulation project, outlined in Chap. 6.

## 3.1 Supervised and Unsupervised Learning

Most machine learning problems can be grouped into two categories: *supervised* and *unsupervised learning*. The majority of practical applications falls into the supervised domain, which deals with function approximation. Given a set of input variables $x_i$



C. Tosciri, *Higgs Boson Decays into a Pair of Bottom Quarks*, Springer Theses,
https://doi.org/10.1007/978-3-030-87938-9_3

with an associated response measurement $y_i$, an algorithm is used to infer the mapping function $f$ from the input to the output:

$$f : X \to Y. \tag{3.1}$$

The set of pairs $\{x_i, y_i\}$, where $i = 1, \ldots, n$, is called a *training set*. In Eq. (3.1), $X$ represents the input space and $Y$ the output space. The goal is to approximate the mapping function in such a way that, given a new input data $x$, it is possible to predict the corresponding new output $y$. A *training process* is required to prepare a model which makes predictions. The training is an iterative process, where the model is dynamically corrected when the prediction is wrong, based on the known output. The training process proceeds until the desired level of *accuracy* is reached. Part of the dataset is generally used to evaluate the performance of the model, it is called a *validation* or a *testing set*.

*Unsupervised learning* is a more challenging task than supervised ones, as in this case for each observable $x_i$ of the dataset, there is not an associated response $y_i$. Nevertheless, it is possible to learn relationships and underlying structures from the dataset and build a model able to predict a new output.

An important aspect of a predictive model is its ability to generalise. A good model must make reasonable predictions on new data. *Overfitting* occurs when the model learns in detail the training dataset, losing the ability to make predictions on new data. On the other hand, *underfitting* refers to a model that can neither model the training data nor generalise to new data. A short training may lead to underfitting the training and testing sets, while too long training can result in overfitting the training dataset and in a poor performance on the testing set. The latter condition is referred to as *overtraining* [6].

## 3.2 Classification and Regression

Supervised learning algorithms can be further categorised into *classification* and *regression* problems. The main difference between the two kinds of tasks is whether the output is from a discrete set or a continuous one. Classification algorithms attempt to estimate the mapping function $f$ from the input variables $x$ to discrete or categorical output $y$. On the other hand, regression algorithms predict continuous output variables. The simplest type of classification is *binary* and allows the prediction of two distinct classes, usually 0 versus 1 or, more common in particle physics, *signal* versus *background*. There is also the possibility to predict more than two class labels with *multi-classification* algorithms and, in particle physics, this is usually applied to distinguish among different kinds of background processes. An example is given by the *boosted decision tree* algorithm (BDT), which is described in the next section.

## *3.2.1 Boosted Decision Tree*

### 3.2.1.1 Multivariate Analyses

A *cut-based* approach is a conventional method to separate signal from background events, by applying a set of cuts on individual kinematic variables characterising the event. Although these observables are chosen to be weakly correlated, a simple cut-based approach is not optimal, since it does not exploit at best the full information carried in the event. To extract results with maximum precision it is necessary to treat the kinematic variables in a *multivariate* way, by optimising multiple selection cuts at a time, in a more sophisticated fashion.

### 3.2.1.2 Decision Tree

One of the most used ML techniques, since decades, is the BDT algorithm [7]. The big success of BDT is due to its simplicity and interpretability. Tree-based models consist in extending a standard cut-based analysis to a multivariate technique. These models can be used for both classification and regression tasks. A simple structure of decision tree is represented in Fig. 3.1.

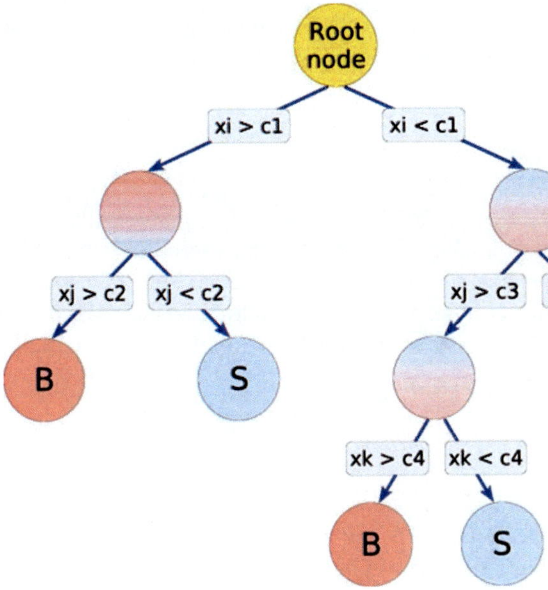


**Fig. 3.1** Schematic structure of a simple decision tree. Starting from the root node, the dataset is split according to a set of decision rules that determine cuts on the discriminating variables $x_i$. Each event of the dataset ends up in one of the leaf nodes at the bottom end of the tree, labelled $S$ for signal and $B$ for background. Image source: [8]

Several observables $x_i$ are fed into the decision tree model which, starting from the *root node*, performs a sequence of binary splits of the data in correspondence of each *node*, according to a set of *decision rules*. The decision rules are determined with specific algorithms that recursively generate the decision tree structure. These algorithms work by choosing a variable at each step that splits the set of items in an optimal way. The separation quality is measured by the specific metric used in the algorithm. One of the most common is the *Gini impurity* or *index* which, in case of many classes, is defined as:

$$\text{Gini} = \sum_{i,j \in \text{classes}}^{i \neq j} p_i p_j, \tag{3.2}$$

where $p_i$ is the probability of assigning a random object to class $i$ and $p_j$ is the probability that the object is actually in class $j$. Gini represents a measure of misclassification or impurity. For a problem with two classes, signal and background, we can define $p$ as the purity of the signal $p = p_s = 1 - p_b$ (where $p_b$ is the purity of background), which is given by:

$$p = \frac{\sum_{i \in \text{signal}} w_s^i}{\sum_{i \in \text{signal}} w_s^i + \sum_{j \in \text{bkg}} w_b^j} = \frac{s}{s+b}, \tag{3.3}$$

where $w_s^i$ and $w_b^j$ are the weights of the signal and background events, respectively, and their corresponding sums are indicated with $s$ and $b$. For a two-class problem, the *Gini* index becomes:

$$\text{Gini} = 1 - \sum_{i=s,b} p_i^2 = 2p(1-p) = \frac{2sb}{(s+b)^2}. \tag{3.4}$$

The goal is to minimise the Gini function, thus the overall tree impurity [9]. This procedure assigns to the node the probability to select an event of a specific class. In the *leaves* of the tree, the terminal nodes, the predictions are based on the probability assigned to each node, according to the distribution of the classes in training data.

#### 3.2.1.3 Boosting

The simple tree-based models are powerful but unstable. A small change in the training data can produce a large change in the tree. This can be mitigated by the use of a *boosting method* [10]: the weights of those training events that have been misclassified are increased (boosted), and a new tree is formed. This procedure is then repeated for the new tree, as many times as the number of boosting stages. This procedure generates several trees which form a *forest* of trees. The new trees are created to correct the residual impurities in the predictions from the existing trees.

From a number of weak classifiers (or *weak learners*), the boosting technique aims at creating stronger classifiers. The whole boosting procedure is controlled by an *objective function* that quantifies the residual impurity and needs to be minimised. The objective function consists in two parts: a *loss function* $L(\theta)$ and a *regularisation* term $\Omega_\theta$:

$$O(\boldsymbol{\theta}) = L(\boldsymbol{\theta}) + \Omega_{\boldsymbol{\theta}}, \tag{3.5}$$

where $\boldsymbol{\theta}$ is the vector of parameters for the model. The loss represents the residual and quantifies the difference between actual value (hypothesis) and predicted value of $y$ (target). A parameter called *learning rate* determines the step size at each boosting iteration, necessary to move toward a minimum of the loss function. Essentially, it controls the speed of learning the corrections. The regularisation measures the complexity of the model. There are many different ways of iteratively adding learners (trees) to minimise an objective function. Some of the most common are the following:

**AdaBoost** Adaptive boosting adjusts the boosting procedure according to the errors returned by the weak learners, by iteratively updating the event weights in a way that favours the events with larger errors. For each input $x$, the outputs of the weak learners are combined as a weighted sum into a function $F(x)$ that represents the model response of the boosted classifier. The algorithm uses the exponential loss function: $L(F, y) = e^{-F(x)y}$ [8, 11].

**GradBoost** Gradient boosting generalise the AdaBoost approach, by using any differentiable loss function, potentially more robust. GradBoost is based on a *gradient descent* procedure which is done by calculating the current gradient of the loss function and then taking steps proportional to the negative of the gradient, in order to reach a minimum of the function [8, 12].

**XGBoost** eXtreme Gradient boosting follows the same principle of gradient boosting. There are, however, differences in the model details. The main improvement is that XGBoost uses regularisation, that helps to penalise some parts of the algorithm and, in general, improves its performance by better controlling overfitting [13].

#### 3.2.1.4 Tree Constraints

The set of inputs is analysed throughout the whole boosted decision tree until a stop condition is verified. The stop conditions are determined by some constraints that can be imposed on the construction of the tree. Below is a list of parameters that can be tuned:

- *Number of Trees* constraints the maximum number of trees in the model.
- *Tree Depth* controls the number of cuts applied before reaching the leaf.
- *Minimal Node Size* imposes a minimum amount of training data at a node before a split can be considered.
- *Minimum Improvement to Loss* quantifies the minimum improvement of any split added to a tree.

A general rule of thumb is that the more strict are the tree conditions, the larger number of trees are needed in the model. In any case, the tree constraints need to be properly tuned in order to construct a good model and avoid overfitting.

## 3.3 Clustering Algorithms

Clustering [1, 10] is a common unsupervised machine learning task, consisting in grouping the dataset into non-overlapping subsets or *clusters*, such that objects within the same cluster are more similar, while objects in different clusters are dissimilar. Sometimes, clustering also involves arranging the clusters into a hierarchy, in such a way that the similarity is kept at each level of the hierarchy. A distance metric between individual objects defines the degree of *similarity*. The choice of the metric depends on the domain of the problem and is critical for catching natural similarity in data. The Euclidean distance is typically employed when quantitative features are used. The Euclidean distance between two vectors **p** and **q** of dimension $d$ is defined as:

$$D(\mathbf{p}, \mathbf{q}) = \sqrt{(q_1 - p_1)^2 + (q_2 - p_2)^2 + \cdots + (q_d - p_d)^2} = \sqrt{\sum_{i=1}^{d} (q_i - p_i)^2}. \tag{3.6}$$

### *3.3.1 K-Means Clustering*

The $K$-means algorithm [14] is one of the most popular clustering methods. It aims at iteratively grouping $N$ observations into a pre-specified number $K$ of clusters in which every observation belongs to the cluster with the nearest mean. The result is a partitioning of the data space into *Voronoi* cells, as indicated in Fig. 3.2, where each panel shows the result of applying $K$-means clustering with a different value of $K$.

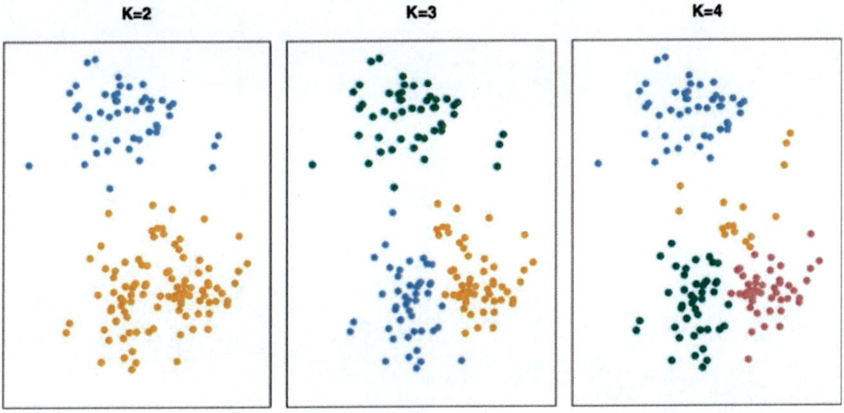


**Fig. 3.2** Results for $K$-means clustering on an unlabelled dataset. Each panel corresponds to a different $K$ used for clustering. The different colours are assigned after the algorithm partitions the dataset. Image source: [10] (Republished with permission of Springer, from "An Introduction to Statistical Learning: with Applications in R", G. James, D. Witten, T. Hastie, R. Tibshirani, 2014; permission conveyed through Copyright Clearance Center, Inc.)

### 3.3.2 *Hierarchical Clustering*

*Hierarchical* clustering [15] is an alternative partitioning approach which does not require to pre-specify the number of clusters $K$ and results in a tree-based representation of the dataset, called *dendrogram*. This is built by at first treating each observation of the dataset as a separate cluster, and then iteratively merging the two most similar clusters, as illustrated in Fig. 3.3. The iteration on which the merging happens, indicated on the vertical axis in Fig. 3.3, measures the degree of similarity between the clusters. Therefore, those observations that are merged at the bottom of the dendrogram are more similar than clusters merged at the top. Cutting the dendrogram at different heights provides different numbers of clusters, the analogous of $K$ in the $K$-means algorithm.

## 3.4 Nearest Neighbours Search

$K$-Nearest Neighbours (KNN) [16] is defined as a *non-parametric*, *lazy* learning algorithm, meaning that it does not make any assumptions on the underlying data distribution (*non-parametric*) and it does not use the training data points to do any generalisation (*lazy*), but it makes predictions from them. Predictions are made for vector $\mathbf{q}$, called *query*, by measuring the distance with all the single instances in the dataset and selecting the $K$ most similar ones. The most popular distance definition is, again, the Euclidean distance, as defined in Eq. (3.6).

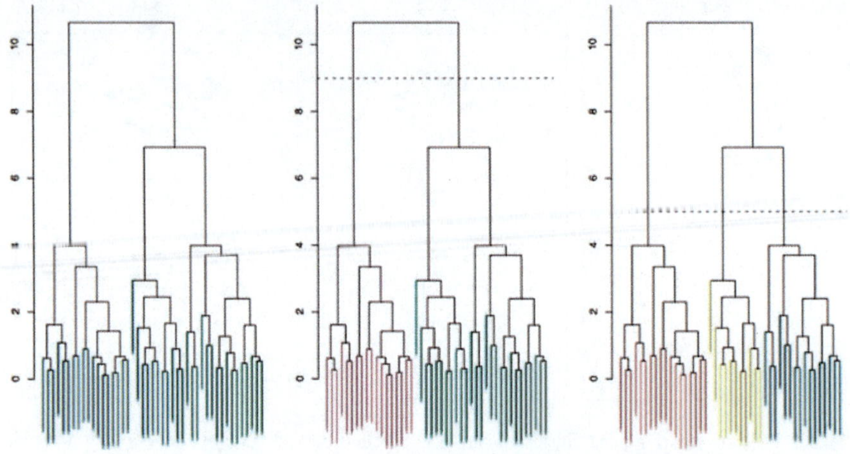

**Fig. 3.3** Dendrogram obtained from hierarchical clustering of a dataset. The different panels show clustering obtained by cutting at different heights. Image source: [10] (Republished with permission of Springer, from "An Introduction to Statistical Learning : with Applications in R", G. James, D. Witten, T. Hastie, R. Tibshirani, 2014; permission conveyed through Copyright Clearance Center, Inc.)

### 3.4.1 Approximate Nearest Neighbours

The computational complexity of KNN scales linearly with the number of instances, making its use difficult for large-scale datasets. Additionally, the volume of the input space increases at an exponential rate with the number of dimensions, leading to the sparsity of the available data. This condition is known as *curse of dimensionality*, which makes the computation unfeasible. To overcome this problem, it is possible to relax the conditions of the exact search of a KNN by permitting a small number of errors. This is achieved by making use of approximate nearest neighbours algorithms [17] that, in some cases, do not guarantee to return exactly the nearest neighbour and result in a less accurate search, but have the advantage to gain in speed and memory saving.

### 3.4.2 Similarity Search

Approximate nearest neighbour search (K-ANNS) [17], or more generally *similarity search*, is receiving growing attention in the current era of *big data* and large information resources. Similarity search is primarily used in computer vision applications, facial recognition and multi-media retrieval to quickly search for objects similar to each other. An exhaustive search for the most similar object is infeasible on large-scale datasets and therefore many approximation methods have been developed. These methods, which are often based on clustering processes, restrict the part of the dataset considered in the search. The dataset restriction involves a pre-processing of the dataset, an operation called *indexing*.

#### 3.4.2.1 Inverted Index

The first indexing approach ever used for billion-scale datasets [18] is based on the *inverted index* structure [19, 20] that requires to split the feature space into Voronoi regions, according to a set of $K$-means centroids, called *codes* and learned on the dataset. Each code is associated to a set of vectors belonging to the corresponding Voronoi cell. The list of vectors is stored in the so-called *inverted file* or *inverted index*. Given a query, the most suitable clusters are identified, based on the similarity with the codes. The partition of the space, as well as the assignment of the query to a cluster, is performed by the *quantiser*, which defines a mapping function between the vectors and the centroids, and in this case is obtained through the $K$-means algorithm. A number of vectors within the selected clusters are returned as a result of the search. An example of the inverted index partitioning is shown in Fig. 3.4 on a dataset of 200 two-dimensional points. The small black dots represent the objects in the dataset, while the green dots are the centroids of the clusters and the big black dot ($\mathbf{q}$) represents the query vector.

#### 3.4.2.2 Product Quantisation

In the case of high dimensionality or when the dataset requires a big amount of computer memory, it can be useful to use a compressed representation of the database vectors. The compressed representations are typically obtained with *product quantisation* (PQ [21]) that allows computing the distance between the query and compressed vectors in a more efficient way. With PQ, each vector is typically split into two sub-vectors, and for each of the two sub-spaces created a dedicated index is built. The quantisation provides a faster search with a less accurate result.

#### 3.4.2.3 Hierarchical Navigable Small World graphs

Proximity graphs have recently gained popularity and are considered state-of-the-art for K-ANNS algorithms [22]. In particular, the model called *Hierarchical Navigable Small World* (HNSW) seems to provide an optimal complexity scaling and good results both for high- and low-dimensional data (with dimension $d \sim 4$) [23]. The idea of the HNSW algorithm is to maintain connectivity between all the vectors of the dataset, through *links*, based on the similarity metric. The graph structure keeps both long-range links to far neighbours and short-range links to closer ones. Links

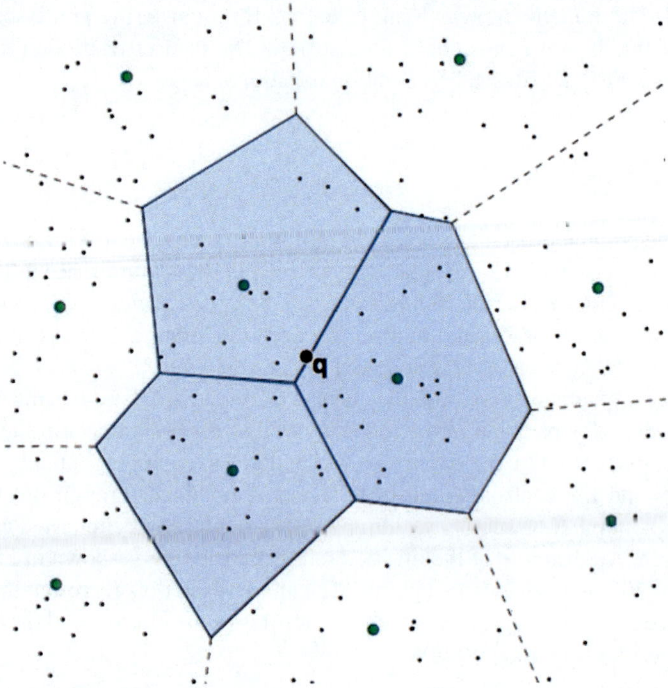


**Fig. 3.4** The Inverted Index partitioning obtained on a dataset of 200 two-dimensional points (small black dots). The three closest cells to the query point (q) are indicated in light blue. Image source: [20] (use permitted under the CC0 1.0 license)

are separated according to their length scale into different layers, with the use of a hierarchical clustering algorithm. A multi-layer structure, similar to the one shown in Fig. 3.5, is then formed.

The first layer contains only a single point, and the base layer is the whole dataset. Each layer contains a so-called *neighbourhood graph*, which connects two points by an edge if there is not a closer third point. The search starts with the first layer. A search is performed on the layer until it reaches the nearest neighbour of the query within this layer. The identified vector is used as an entry point to perform the search in the next layer. The procedure is repeated until the bottom layer is reached. In this way only a needed fixed portion of the connections for each element is evaluated, thus allowing a logarithmic scalability. HNSW is expected to provide an exceptional trade-off between speed and accuracy.

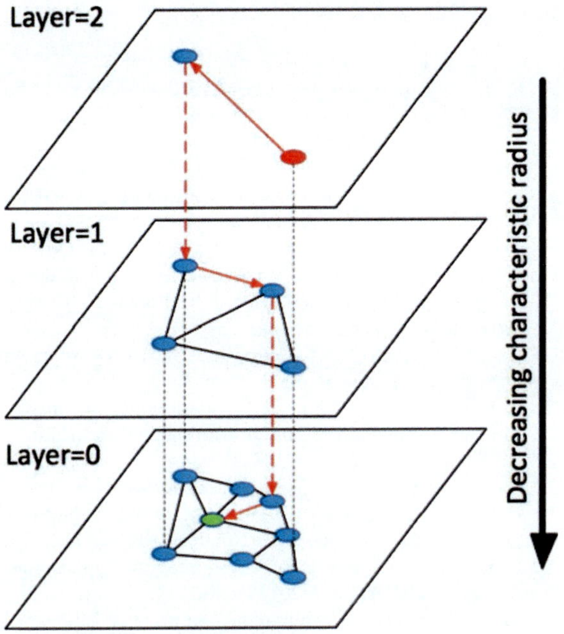


**Fig. 3.5** Illustration of an HNSW structure composed by three layers. Image source: [23] (use permitted under the CC0 1.0 license)

## 3.5 Machine Learning in High Energy Physics

Recently, ML methods have made a prominent appearance in the field of particle physics. Although simple implementations of neural networks [5] and BDTs have been used for decades to improve the discrimination between signal and background, and consequently the sensitivity of physics analyses, the deep learning revolution occurring in the industrial field is also reaching the HEP community.

In the next decade, the LHC machine will undergo major upgrades to guarantee sufficient statistic for performing precise measurements. The set of upgrades will enhance the sensitivity to new physics by progressively increasing the *pp* collision rate. The data volume, size and complexity will bring a variety of technological challenges, such as real-time particle reconstruction and fast detector simulation, that the particle physics community will need to address already in the Run-3 of the LHC (2022–2025). The qualitative and quantitative complexity of the data collected in the future at the LHC will pose limits to the physics performance of algorithms and computational resources. ML methods, designed to exploit large datasets to reduce complexity and find new features in data, promise to provide improvements in both of these areas. ML techniques could outperform traditional and resource-consuming techniques in tasks typical of physics experiments at particle colliders, such as particle identification and reconstruction, energy measurement, and detector simulation.

## References

1. Hastie T, Tibshirani R, Friedman JH (2009) The elements of statistical learning: data mining, inference, and prediction. Springer series in statistics. Springer. https://web.stanford.edu/~hastie/ElemStatLearn/
2. Radovic A, et al (2018) Machine learning at the energy and intensity frontiers of particle physics. Nature 560. https://doi.org/10.1038/s41586-018-0361-2
3. Carleo G, et al (2019) Machine learning and the physical sciences. Rev Mod Phys 91(4):045002. https://doi.org/10.1103/RevModPhys.91.045002. arXiv: 1903.10563 [physics.comp-ph]
4. Albertsson K, et al (2018) Machine learning in high energy physics community white paper. J Phys Conf Ser 1085(2):022008. https://doi.org/10.1088/1742-6596/1085/2/022008. arXiv: 1807.02876 [physics.comp-ph]
5. Goodfellow I, Bengio Y, Courville A (2016) Deep learning. MIT Press, Cambridge. http://www.deeplearningbook.org
6. Brownlee J, Overfitting and underfitting with machine learning algorithms. https://machinelearningmastery.com/overfitting-and-underfitting-with-machine-learning-algorithms/
7. Breiman L, et al (1984) Classification and regression trees. Wadsworth and Brooks, Monterey
8. Hocker A, et al (2007) TMVA - toolkit for multivariate data analysis with ROOT: users guide. Technical report, Geneva: CERN. https://cds.cern.ch/record/1019880
9. Coadou Y (2013) Boosted decision trees and applications. EPJ Web Conf 55:02004. https://doi.org/10.1051/epjconf/20135502004
10. James G, et al (2014) An introduction to statistical learning: with applications in R. Springer Publishing Company, Incorporated. ISBN: 1461471370
11. Freund Y, Schapire RE (1997) A decision-theoretic generalization of on-line learning and an application to boosting. J Comput Syst Sci 55(1):119–139. ISSN: 0022-0000. https://doi.org/10.1006/jcss.1997.1504. http://www.sciencedirect.com/science/article/pii/S002200009791504X
12. Friedman J (2000) Greedy function approximation: a gradient boosting machine. Ann Stat 29. https://doi.org/10.1214/aos/1013203451
13. Chen T, Guestrin C (2016) XGBoost: a scalable tree boosting system. arXiv:1603.02754
14. MacQueen J (1967) Some methods for classification and analysis of multivariate observa- tions. In: Proceedings of the fifth berkeley symposium on mathematical statistics and probability, vol 1: Statistics. University of California Press, Berkeley, pp 281–297. https://projecteuclid.org/euclid.bsmsp/1200512992
15. Jardine N, van Rijsbergen CJ (1971) The use of hierarchic clustering in information retrieval. Inf Storage Retr 7(5):217–240. ISSN: 0020- 0271. https://doi.org/10.1016/0020-0271(71)90051-9. http://www.sciencedirect.com/science/article/pii/0020027171900519
16. Cunningham P, Delany S (2007) k-Nearest neighbour classifiers. Mult Classif Syst
17. Indyk P, Motwani R (2000) Approximate nearest neighbors: towards removing the curse of dimensionality. In: Conference proceedings of the annual ACM symposium on theory of computing, pp 604–613. https://doi.org/10.1145/276698.276876
18. Jégou H, et al (2011) Searching in one billion vectors: re-rank with source coding. In: ICASSP 2011 - International conference on acoustics, speech and signal processing. Prague, Czech Republic: IEEE, May 2011, pp 861–864. https://doi.org/10.1109/ICASSP.2011.5946540. https://hal.inria.fr/inria-00566883
19. Babenko A, Lempitsky V (2015) The inverted multi-index. IEEE Trans Pattern Anal Mach Intell 37(6):1247–1260. ISSN: 1939-3539. https://doi.org/10.1109/TPAMI.2014.2361319
20. Baranchuk D, Babenko A, Malkov Y (2018) Revisiting the inverted indices for billion-scale approximate nearest neighbors. arXiv:1802.02422
21. Jégou H, Douze M, Schmid C (2011) Product quantization for nearest neighbor search. In: IEEE transactions on pattern analysis and machine intelligence 33(1):117-128. ISSN: 1939-3539. https://doi.org/10.1109/TPAMI.2010.57

22. Douze M, Sablayrolles A, Jégou H (2018) Link and code: fast indexing with graphs and compact regression codes. arXiv:804.09996
23. Malkov YA, Yashunin DA (2016) Efficient and robust approximate nearest neighbour search using Hierarchical Navigable Small World graphs. arXiv: 1603.09320

# Chapter 4
# The Large Hadron Collider and the ATLAS Detector

## 4.1 The Large Hadron Collider

The Large Hadron Collider (LHC) is the largest and most powerful particle accelerator ever built for particle physics research, operating at the European Organization for Nuclear Research (CERN) near Geneva, in Switzerland. The collider is installed in the tunnel of 27 km of circumference that hosted the former Large Electron-Positron Collider (LEP) [1] and is located 100 m underground. LHC is a two-ring-superconducting-hadron synchrotron designed to accelerate bunches of protons in opposite directions and collide them up to a centre-of-mass energy of $\sqrt{s} = 13$ TeV. The machine is also capable of accelerating heavy ions. The analyses and studies reported on this thesis are obtained with $pp$ collisions at $\sqrt{s} = 13$ TeV.

### *4.1.1 The Accelerator Complex*

Before being injected in the LHC ring and collided, the protons are progressively accelerated through the pre-accelerator chain, as shown in Fig. 4.1. The proton source originates from the ionization of pure hydrogen gas through the use of an electric field. The proton beam is collected and accelerated up to 50 MeV through the Linear Accelerator 2 (LINAC2) and then injected into the Proton Synchrotron Booster (PSB). Here, protons are accelerated to 1.4 GeV, before their energy is boosted by the Proton Synchrotron (PS), up to 25 GeV. The Super Proton Synchrotron (SPS) accelerates them to 450 GeV. Then, the protons are split and they finally reach the two beam-pipes of the LHC ring, where they circulate in opposite directions for about 20 minutes before gaining the final energy of 6.5 TeV. Protons are aggregated in bunches containing $1.1 \times 10^{11}$ protons and separated from each other by a 25 ns interval. The two beams are brought into collision in four interaction points, within four detectors: ATLAS [3], ALICE [4], CMS [5] and LHCb [6].



C. Tosciri, *Higgs Boson Decays into a Pair of Bottom Quarks*, Springer Theses,
https://doi.org/10.1007/978-3-030-87938-9_4

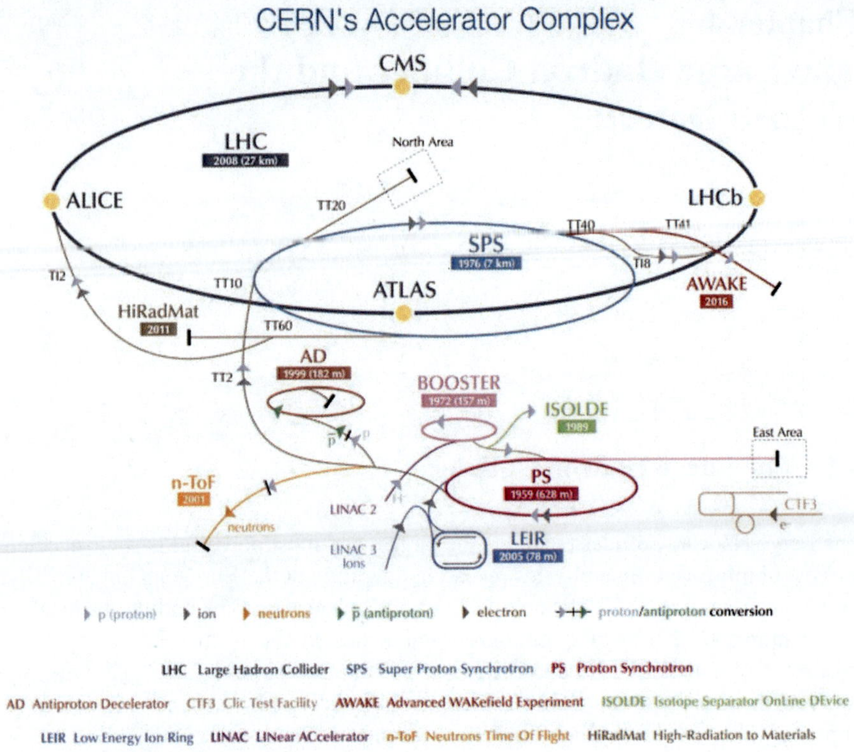


**Fig. 4.1** The CERN accelerator complex. *Image source* [2]

### 4.1.2 Luminosity

The capability of a collider is measured by the *luminosity* $\mathcal{L}$, the instantaneous rate of beams interaction per cross section $\sigma$ of the *pp* process, given by

$$\frac{dN}{dt} = \mathcal{L} \cdot \sigma. \tag{4.1}$$

The instantaneous luminosity can be defined as a function of the beams parameters:

$$\mathcal{L} = \frac{N_1 N_2 f N_b}{4\pi \sigma_x \sigma_y}, \tag{4.2}$$

where $N_{1(2)}$ are the intensities of two colliding bunches, $f$ is the revolution frequency, $N_b$ is the number of bunches in one beam and $\sigma_{x(y)}$ refers to the beam width, assuming two Gaussian beams colliding head-on. An important parameter is given by the integrated luminosity over a certain period of data taking $L = \int \mathcal{L}\, dt$, which quantifies the performance of an accelerator. In Fig. 4.2 is shown the integrated luminosity

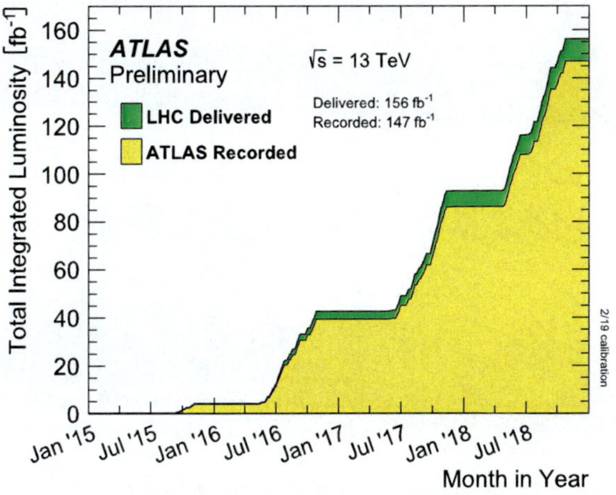


**Fig. 4.2** Integrated luminosity versus time delivered by the LHC (green) and collected by ATLAS (yellow) during stable beams for $pp$ collisions at $\sqrt{s} = 13$ TeV in Run-2. *Image source* [7]

delivered by the LHC and collected by the ATLAS experiment, during the period of data-taking between 2015 and 2018 for collisions at 13 TeV centre-of-mass energy (Run-2).

At the LHC, an *event* is identified as a head-on collision occurring between two protons within a bunch crossing and producing the hard scattering of interest. These head-on collisions, however, are always accompanied by multiple soft proton-proton collisions in the same and adjacent beam crossing, called *pile-up*. As the instantaneous luminosity increases, the number of additional interactions rises. The number of interactions per crossing is calculated for each bunch as

$$\mu = \frac{\mathcal{L}_{\text{bunch}}\, \sigma_{\text{inel}}}{N_b\, f}, \tag{4.3}$$

where $\mathcal{L}_{\text{bunch}}$ is the instantaneous luminosity per bunch and $\sigma_{\text{inel}}$ is the inelastic $pp$ cross section. The mean number of interactions per crossing corresponds to the mean of the Poisson distribution of the number of interactions per crossing calculated for each bunch. This distribution is shown in Fig. 4.3.

### 4.1.3 The LHC Program

The LHC operations started in 2009 with collisions at a centre-of-mass energy of $\sqrt{s} = 7$ TeV followed in 2012 by collisions at $\sqrt{s} = 8$ TeV. The data collected between 2010 and 2012 are referred to as Run-1. The LHC machine and the detectors were repaired and upgraded during a Long Shutdown (LS1) between 2013 and 2015. In 2015, operations resumed for Run-2 at $\sqrt{s} = 13$ TeV and the machine progressively increased the delivered luminosity until the end of 2018, when another

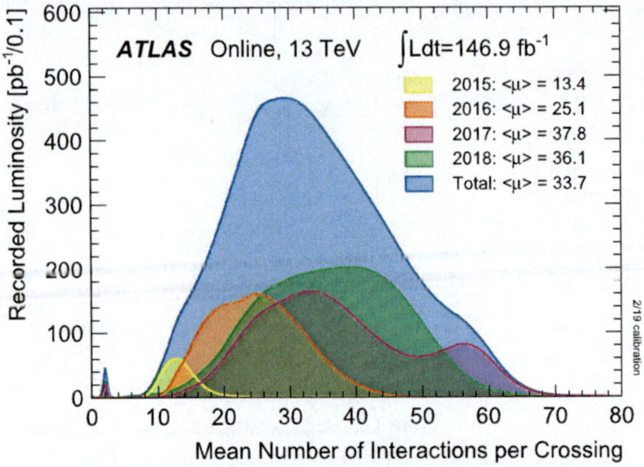


**Fig. 4.3** Luminosity-weighted distribution of the mean number of interactions per crossing for 2015 (yellow), 2016 (orange), 2017 (purple), 2018 (green) and total (blue), obtained with data of $pp$ collisions at $\sqrt{s} = 13$ TeV. *Image source* [7]

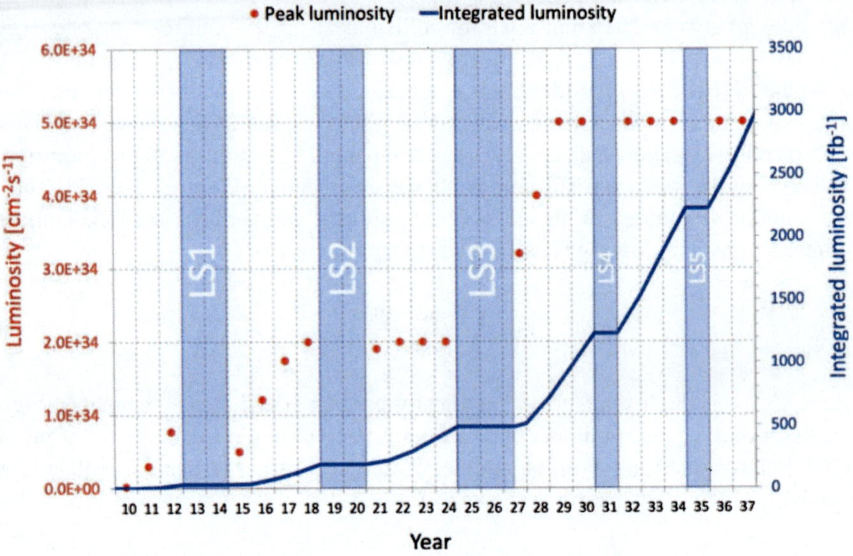


**Fig. 4.4** Projected LHC performance through 2037, showing preliminary dates for long shutdowns of the LHC and projected luminosities. *Image source* [8]

shutdown, LS2, started for maintenance and upgrades. In Fig. 4.4, a sketch of the LHC tentative plan is reported. After Run-3, scheduled to start in 2022, a third LS will precede the High Luminosity configuration (HL-LHC). During LS3 the machine will undergo major upgrades to increase its luminosity by a factor of at least five beyond its nominal design value. The significant increase of luminosity is necessary to guarantee sufficient statistic for performing precise measurements.

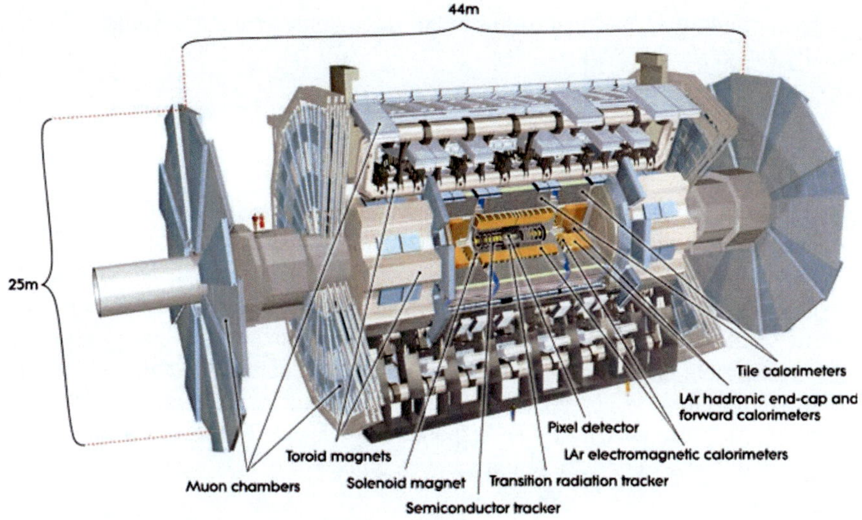


**Fig. 4.5** The ATLAS detector. The main sub-systems are indicated. *Image source* [9]

## 4.2 The ATLAS Detector

The ATLAS (A Toroidal LHC ApparatuS) [3] is one of the main experiments located at the LHC. It is a general-purpose detector devoted to the study of a wide range of physics processes, in particular to measure precisely the SM parameters and to search for BSM particles. The ATLAS detector, shown in Fig. 4.5, surrounds one of the crossing points of the beams and is shaped like a cylinder. It consists of a set of detectors arranged in layers to cover almost the entire solid angle and detect all the particles produced in the collisions. The ATLAS detector is forward-backward symmetric to the interaction point.

### *4.2.1 Coordinate System*

The ATLAS reference frame is a right-handed Cartesian coordinate system, where the origin is identified with the nominal interaction point, the $z$-axis lies on the beam direction and the $x - y$ plane is transverse to the beam direction. The positive $x$-axis points to the centre of the LHC ring and the positive $y$-axis points upwards. The azimuthal angle $\phi$ is defined with respect to the $x$-axis on the $x - y$ plane, whilst $\theta$ is measured on the $y - z$ plane and with respect to the $z$-axis. However, to indicate the polar direction it is more convenient to use the *pseudorapidity*, which is defined as:

$$\eta = -\ln \tan\left(\frac{\theta}{2}\right). \tag{4.4}$$

The pseudorapidity is used in the ultra-relativistic regime as an approximation of the quantity called *rapidity*, defined by:

$$y = \frac{1}{2} \ln \left( \frac{E + p_z}{E - p_z} \right), \tag{4.5}$$

where $p_z$ is the component of the momentum along the beam axis. The advantage of using these quantities is that their intervals are invariant under a boost along the longitudinal direction. The angular distance $\Delta R$ between two objects can be conveniently defined as $\Delta R = \sqrt{\Delta\phi^2 + \Delta\eta^2}$. Other important observables for the description of interesting events detected with ATLAS are the transverse momentum $p_T = p \cos\theta$ and the transverse energy $E_T = E \cos\theta$, defined on the $x - y$ plane. The missing transverse energy $E_{\mathrm{T}}^{\mathrm{miss}}$ quantifies the energy that is not detected on the transverse plane and mostly corresponds to undetected neutrinos in the event. According to energy conservation, it can be computed by balancing the transverse energy detected, as described in Section 5.7.

### 4.2.2 *Inner Detector*

The ATLAS Inner Detector (ID) [10, 11] is used to track charged particles from the interaction region to the electromagnetic calorimeter system and measure their momentum within the pseudorapidity range $|\eta| < 2.5$. The ID operates in an environment characterised by high track density. The detector fine granularity provides the high resolution required for the momentum and vertex reconstruction. The ID consists of three different systems of sensors, all immersed in a 2T magnetic field parallel to the beam axis, generated by a superconducting solenoid which surrounds it. The layout of the ID is illustrated in Fig. 4.6.

**Pixel Detector (Pixels)** The innermost system is composed of silicon pixel detectors. It originally consisted of three barrel layers and three disks on each detector side, for a total of about 80 million pixels and 1744 readout modules. Pixels have a typical size of 50 $\mu$m $\times$ 400 $\mu$m and a thickness of 250 $\mu$m [12]. During the LS1, the barrel was supplemented with a fourth layer, called Insertable B-layer (IBL) [13]. Located at a distance of 33 mm to the beamline, it contains 12 million pixels with size 50 $\mu$m $\times$ 250 $\mu$m. The IBL was introduced to improve the tracking robustness and precision. These are important requirements for the determination of the track impact parameter, fundamental for the performance of $b$-tagging, described in Section 5.5.4.

**SemiConductor Tracker (SCT)** The sub-detector that surrounds the Pixels consists of silicon microstrips sensors assembled in four concentric barrels and two end-caps composed by nine disks each. Every component, either barrel or disk, provides two measurements which are combined in a single space-point. Every particle originated in the interaction point is typically associated with eight strip measurements (hits) which provide four space-point measurements [14]. The barrel sector has 2112 mod-

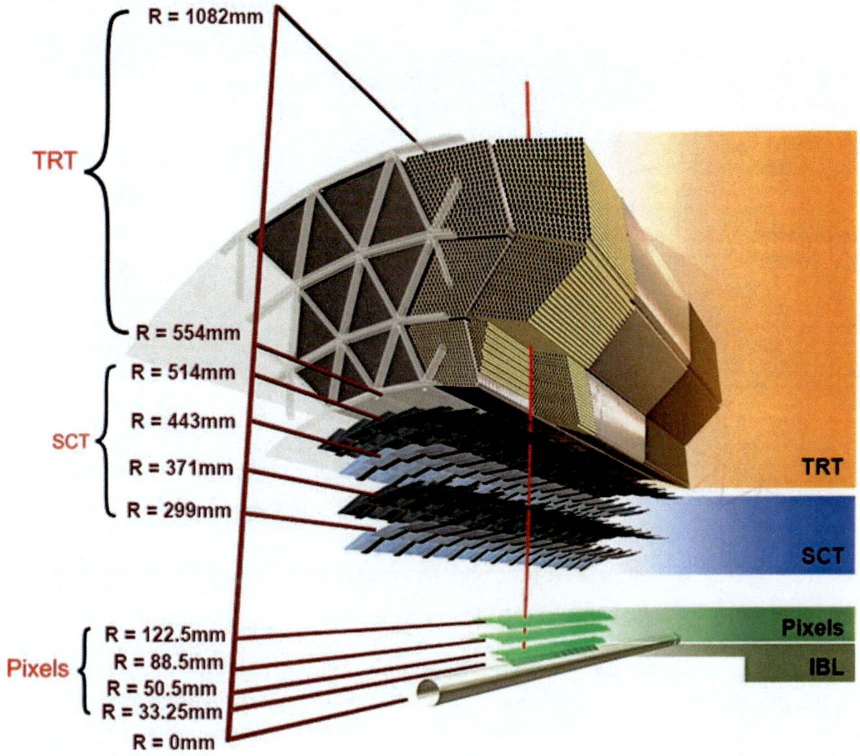


**Fig. 4.6** Cut-away view of the ATLAS inner detector. *Image source* [12]

ules, while each end-cap counts 988 modules. Each barrel module consists of four silicon-strip sensors with strips parallel to the beamline at a pitch of 80 $\mu$m. Two rectangular sensors are paired together and contain a total of 768 strips of approximately 12 cm in length, which provide 2D measurements. Another pair is glued back-to-back to the first one at a stereo angle of 40 mrad, to provide information on the third coordinate. The modules on the end-caps have a similar arrangement with trapezoidal sensors characterised by strip pitch of variable size (80 $\mu$m in average) and mounted in such a way that the strips have radial direction. The thickness of the sensors is 285 $\mu$m.

**Transition Radiation Tracker (TRT)** The TRT [15] is the outermost part of the ID. It is a straw-tube tracker which contains around 300 thousands cylindrical drift tubes (straws) with a diameter of 4 mm, arranged in parallel to the beam axis and serving as cathodes. Each tube contains a gold-plated tungsten wire in its centre and is filled with a gas mixture based on xenon. When a charged particle traverses a straw, it ionises the gas and produces electrons that drift toward the wire where they are collected and read out. The tubes are embedded in polymer fibres and foils which provide a transition material boundary. In traversing the boundary, a

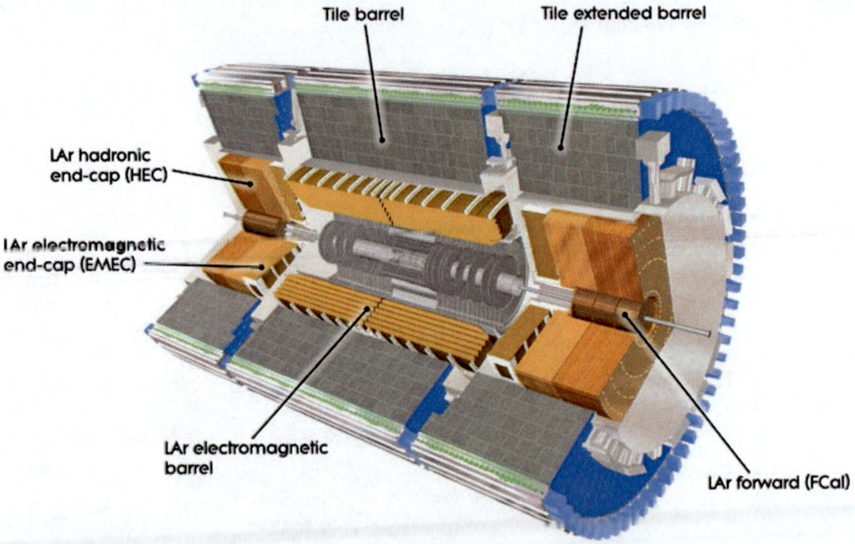


**Fig. 4.7** Cut-away view of the ATLAS calorimeter system. *Image source* [18]

relativistic charged particle may emit radiation with a probability that depends on the relativistic factor $\gamma = E/m$. The transition radiation photons are absorbed in the straw, producing high-threshold signals. Since the emission is much more likely for an electron than for a hadronic particle, such as a pion, it is more probable that an electron track is associated with higher energy hits, providing a good discrimination power for particle identification. The intrinsic resolution of the TRT is larger than in the silicon trackers and corresponds to $120\,\mu$m. However, the TRT provides a large number of hits per track, typically $> 30$. This design makes the TRT an optimal detector, not only for the capability of particle identification but also for good pattern recognition.

### 4.2.3 Calorimetry

The calorimetry in ATLAS [16] consists of a set of sampling detectors with full symmetry in $\phi$ and coverage around the beam axis, as shown in Fig. 4.7. The system is designed to stop or absorb most of the particles coming from a collision and to measure their energy. In most cases, the Liquid Argon (LAr) technology [17], consisting in the use of LAr as active detector material, is employed. The advantages of using LAr rely on the stability of its response over time and its intrinsic radiation-hardness. Electromagnetic calorimeters, characterised by a fine granularity, are designed to precisely measure the energy loss of electrons and photons, through the electromagnetic showers created when they traverse the medium. The development of the showers is parameterised by the *radiation length* ($X_0$), the average distance over which a high energy electron is reduced to $1/e$ of its energy by *bremsstrahlung*. The depth of the electromagnetic calorimeters is measured in radiation lengths and, in ATLAS, it is larger than 22 $X_0$. Similarly, the evolution of hadronic showers in the matter, which is mainly governed by ionisation and nuclear interactions, is characterised by the *nuclear interaction length* λ, defined as the mean distance travelled by a hadron, before interacting with a nucleus. The depth of the hadronic calorimeters, which require coarser granularity to measure hadronic jets and $E_{\mathrm{T}}^{\mathrm{miss}}$, is $\sim 10$ λ. These dimensions provide good containment for the particle showers.

The energy resolution of a calorimeter is described by

$$\frac{\sigma_E}{E} \approx \frac{a}{\sqrt{E}} \oplus \frac{b}{E} \oplus c, \tag{4.6}$$

where the parameter $a$ is called *stochastic term* and represents the statistical shower development, $b$ is the *noise term*, $c$ is the *constant term* and the symbol $\oplus$ indicates a quadratic sum. The nominal energy resolution for the electromagnetic calorimeter is $\frac{\sigma_E}{E} \approx \frac{10\%}{\sqrt{E}} \oplus 0.7\%$, while for the hadronic calorimeter is $\frac{\sigma_E}{E} \approx \frac{50\%}{\sqrt{E}} \oplus 3\%$ [19].

**Electromagnetic Calorimeters (ECAL)** The ECAL is a lead-LAr detector [17] consisting of a barrel component (EMB) which covers the pseudorapidity region of $|\eta| < 1.475$ and two end-cap components (EMEC), covering $1.375 < |\eta| < 3.2$. The lead plates and electrodes are arranged in an accordion geometry, with size varying with $\eta$, to guarantee complete $\phi$ symmetry without azimuthal cracks.

The EMB is composed of three layers with different thickness and segmentation in $\eta$. The first layer has the finest granularity in $\eta$ to provide $\gamma - \pi^0$ separation and photon direction measurement. The second and third layers measure most of the electromagnetic shower energy and leakage, respectively. The three layers are complemented by a presampler, an instrumented argon layer, which provides a measurement of the energy lost in front of the electromagnetic calorimeters.

The EMEC calorimeter consists of two coaxial wheels, one on each side of the electromagnetic barrel. It presents a longitudinal segmentation and its thickness is similar to that of the EMB.

**Hadronic Calorimeters (HCAL)** The HCAL is located outside the ECAL envelope and consists of sampling calorimeters which employ different technologies, according to the physics performance required.

The Tile calorimeter [20] uses steel as the absorber and plastic scintillating tiles as the active material. The tiles are oriented radially and perpendicular to the beam-pipe and are read out by wave-length shifting fibres, which deliver the light to photomultipliers. The Tile calorimeter is composed by a barrel in the range $|\eta| < 1.0$ and two extended barrels in $0.8 < |\eta| < 1.7$.

The Hadronic End-cap Calorimeter (HEC) is a sampling calorimeter made with copper plates interleaved with 2 mm gaps filled with LAr. The HEC consists of two independent wheels per end-cap, located directly behind the EMEC, and covering

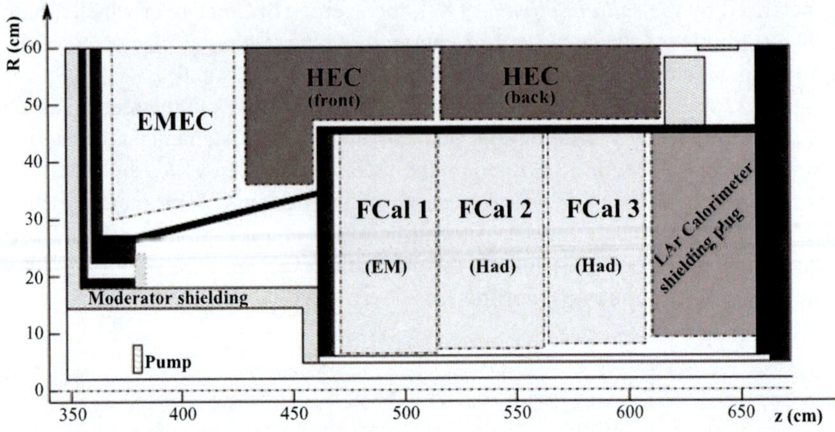


**Fig. 4.8** Transversal view of the ATLAS calorimeter system showing the transition from the EMEC and HEC calorimeters to the FCAL calorimeter. *Image source* [23]

the range $1.5 < |\eta| < 3.2$. Each wheel is divided into two segments in depth, for a total of four layers per end-cap.

### *4.2.4 Forward Calorimeter*

The most forward region of the ATLAS detector is equipped with a forward calorimeter (FCal) [21, 22], which extends the pseudorapidity coverage, spanning $3.1 < |\eta| < 4.9$ and slightly overlapping with the HEC, as shown in Fig. 4.8. The FCal, located at a distance of $\sim$ 4.7 m to the nominal interaction point, completes the nearly hermetic calorimeter system in ATLAS. It is designed to detect jets originated from high-rate low-$p_{\text{T}}$ or soft *minimum-bias* collisions which produce particles whose density and energy increase with $\eta$. These jets would otherwise escape detection, leading to false $E_{\text{T}}^{\text{miss}}$ signatures.

Due to the harsh environment near the beamline, the FCal, which is a LAr sampling calorimeter, is designed with an unconventional structure, as shown in the front-face view in Fig. 4.9. The FCal consists of an absorber matrix which holds cylindrical electrodes arranged parallel to the beamline. Each electrode consists of a *rod* (anode) inside a *tube* (cathode), separated by a small gap (0.27 mm) filled with LAr, the active medium. In each end-cap, the FCal is segmented into three modules. The first module (FCal1), closest to the interaction point, uses a copper absorber and is optimised for electromagnetic measurements. The following modules (FCal2 and FCal3) are made with tungsten and are used as hadronic calorimeters and to correct for leakage behind the electromagnetic calorimeter. In the FCal1 module, two consecutive electrodes are separated on the $x$-axis by a distance of 7.5 mm. This distance, as well as the LAr

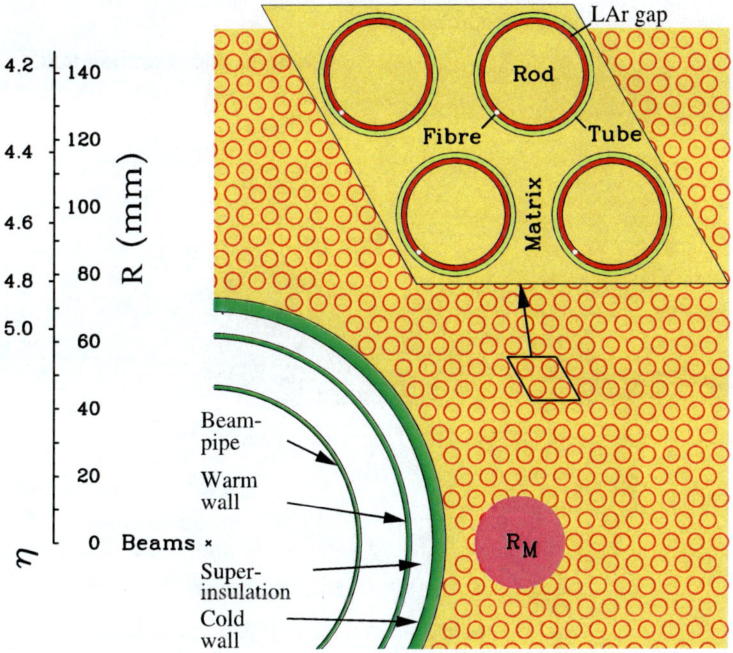


**Fig. 4.9** Front-face view of the electrode arrangement of a portion of the forward calorimeter (FCal) in ATLAS. *Image source* [23]

gap, increases in the FCal2 and FCal3 modules. The nominal energy resolution for the FCal is $\frac{\sigma_E}{E} \approx \frac{100\%}{\sqrt{E}} \oplus 10\%$

### 4.2.5 Muon Spectrometer

The muon spectrometer (MS) [24, 25], shown in Fig. 4.10, is the outermost sub-detector in ATLAS dedicated to measure the trajectory and momentum of the charged particles exiting the calorimeters, typically muons. The MS is based on a magnetic system of three large air-core superconducting toroidal magnets, one barrel and two end-caps, each consisting of eight flat coils. The magnetic system provides a field of approximately 0.5 T in the region of pseudorapidity $|\eta| < 2.7$ which deflects the charged particles entering the MS. The transverse momentum of the muons is measured with a resolution that varies between 3% (for muons with $2 < p_{\mathrm{T}} < 250$ GeV) and 10% (for muons with $p_{\mathrm{T}} \approx 1$ TeV). The MS is instrumented with two separate kind of chambers, optimised either for trigger or high-precision tracking.

**Precision Tracking Chambers** Precise measurements of the muon trajectory deflection in the barrel are provided by the Monitored Drift Tubes (MDT) [27], arranged in

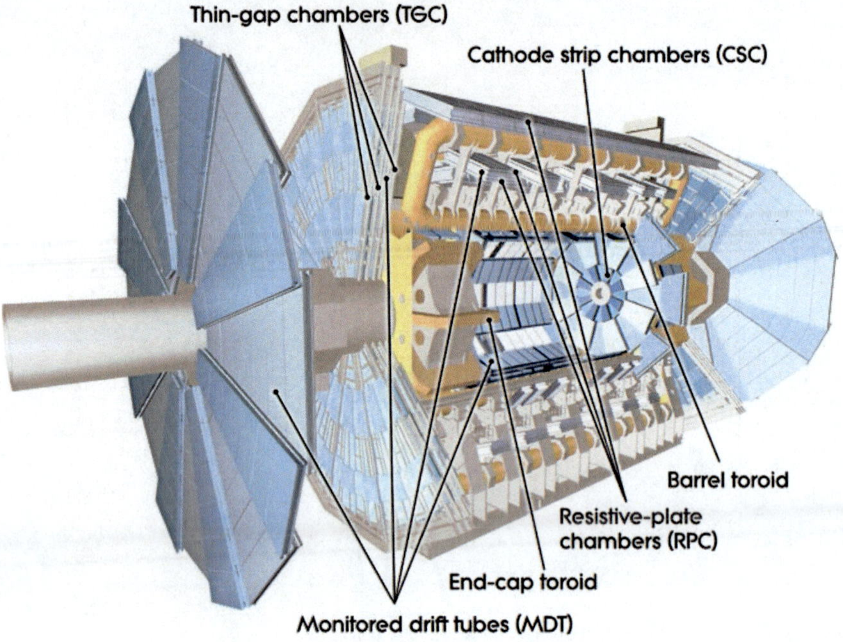


**Fig. 4.10** Cut-away view of the ATLAS muon spectrometer. *Image source* [26]

three or four layers and covering the region of pseudorapidity $|\eta| < 2.7$. The MDT are designed to provide a track position resolution of $40\mu$m. The region of pseudorapidities of $2.0 < |\eta| < 2.7$ is covered by Cathode Strip Chambers (CSC) [28], multi-wire proportional chambers with cathodes segmented into strips.

**Fast Trigger Chambers** The trigger system covers the pseudorapidity range $|\eta| < 2.4$ and consists of three layers of Resistive Plate Chambers (RPC) [29], which are used in the barrel, and three or four layers of Thin Gap Chambers (TGC) [30], used in the end-cap regions. The main purpose of this system is to provide the muon hardware-based trigger, through the hit coincidences of different layers. Furthermore, the system provides a measurement of the muon trajectory in the direction orthogonal to the precision measurement that is approximately parallel to the magnetic field lines.

### 4.2.6 *Trigger and Data Acquisition*

During Run-2, the LHC bunch-crossing occurred with a frequency of 40 MHz, resulting in a large and complex data flow. Due to the limited bandwidth in the online and offline computing resources, the recording rate must be reduced by rejecting minimum-bias processes (produced without any hard collisions occurring in the event), while maintaining efficiency in selecting relevant physics processes.

The trigger system in ATLAS [31] is responsible for retaining the most interesting events for later study, reducing the recorded event rate to 1 kHz. The trigger system consists of a hardware-based Level-1 (L1) trigger and a software-based high-level trigger (HLT). The L1 trigger uses information from the calorimeters and the muon spectrometer to identify signatures from high-$p_{\mathrm{T}}$ objects. It also targets events with large $E_{\mathrm{T}}^{\mathrm{miss}}$ and $E_{\mathrm{T}}$. The L1 reduces the event rate to 100 kHz. The $\eta - \phi$ regions of the detector where interesting features have been identified, are defined as Regions-of-Interest (RoI) and passed to the HLT. The HLT performs an event reconstruction by executing offline-like algorithms within the RoI.

A trigger menu defines the selection criteria for L1 and HLT. The trigger configuration depends on the purpose of the data taking. Different triggers are used for physics measurements, efficiency and performance measurements and detector calibration.

## References

1. Myers S, Picasso E (1990) The design, construction and commissioning of the CERN large Electron-Positron collider. In: Contemporary physics, vol 31.6, pp 387–403. https://doi.org/10.1080/00107519008213789
2. Marcastel F (2013) CERN's accelerator complex. La chaîne des accélérateurs du CERN. In: General photo, Oct 2013. https://cds.cern.ch/record/1621583
3. The ATLAS Collaboration (2008) The ATLAS experiment at the CERN large hadron collider. J Instrum 3.08:S08003–S08003. https://doi.org/10.1088/1748-0221/3/08/s08003. https://doi.org/10.1088%2F1748-0221%2F3%2F08%2Fs08003
4. The ALICE Collaboration et al (2008) The ALICE experiment at the CERN LHC. J Instrum 3.08:S08002–S08002. https://doi.org/10.1088/1748-0221/3/08/s08002. https://doi.org/10.1088%2F1748-0221%2F3%2F08%2Fs08002
5. The CMS Collaboration et al (2008) The CMS experiment at the CERN LHC. J Instrum 3.08:S08004–S08004. https://doi.org/10.1088/1748-0221/3/08/s08004. https://doi.org/10.1088%2F1748-0221%2F3%2F08%2Fs08004
6. The LHCb Collaboration et al (2008) The LHCb Detector at the LHC. J Instrum 3.08:S08005–S08005. https://doi.org/10.1088/1748-0221/3/08/s08005. https://doi.org/10.1088%2F1748-0221%2F3%2F08%2Fs08005
7. ATLAS Collaboration (2019) Luminosity Public Results Run-2. July 2019. https://twiki.cern.ch/twiki/bin/view/AtlasPublic/LuminosityPublicResultsRun2
8. Schedules and luminosity forecasts. https://lhc-commissioning.web.cern.ch/lhc-commissioning/schedule/HL-LHC-plots.htm
9. Pequenao J (2008) Computer generated image of the whole ATLAS detector, Mar 2008. https://cds.cern.ch/record/1095924
10. ATLAS inner detector (1997) Technical Design Report, 1. Technical Design Report ATLAS. CERN, Geneva. http://cds.cern.ch/record/331063
11. Stanecka E (2013) The ATLAS inner detector operation, data quality and tracking per- formance. In: Proceedings of 32nd international symposium on physics in collision (PIC 2012): Strbske Pleso, Slovakia, 12–15 Sep 2012, pp 383–388. arXiv: 1303.3630 [physics.ins-det]
12. Potamianos K (2015) The upgraded Pixel detector and the commissioning of the Inner Detector tracking of the ATLAS experiment for Run-2 at the Large Hadron Collider. Technical report ATL-PHYS-PROC-2016-104. 15 pages, EPS-HEP 2015 Proceedings. CERN, Geneva, Aug 2016. https://cds.cern.ch/record/2209070

13. Capeans M et al (2010) ATLAS Insertable B-Layer technical design report. Technical report CERN- LHCC-2010-013. ATLAS-TDR-19. Sept 2010. https://cds.cern.ch/record/1291633
14. Aad G et al (2014) Operation and performance of the ATLAS semiconductor tracker. In: JINST 9:P08009. https://doi.org/10.1088/1748-0221/9/08/P08009. arXiv:1404.7473 [hep-ex]
15. Vogel A (2013) ATLAS transition radiation tracker (TRT): straw tube gaseous detectors at high rates. Technical report ATL-INDET-PROC-2013-005. CERN, Geneva, Apr 2013. https://cds.cern.ch/record/1537991
16. Airapetian A et al (1999) ATLAS: detector and physics performance technical design report, vol 1
17. ATLAS liquid-argon calorimeter (1996) Technical Design Report. Technical Design Report ATLAS. CERN, Geneva. https://cds.cern.ch/record/331061
18. Pequenao J (2008) Computer generated image of the ATLAS calorimeter, Mar 2008. https://cds.cern.ch/record/1095927
19. Ilic N (2014) Performance of the ATLAS Liquid Argon Calorimeter after three years of LHC operation and plans for a future upgrade ATLAS Liquid Argon Calorimeter after three years of LHC operation and plans for a future upgrade. J Instrum 9.03:C03049–C03049. https://doi.org/10.1088/1748-0221/9/03/c03049. https://doi.org/10.1088%2F1748-0221%2F9%2F03%2Fc03049
20. Correia AMH (2015) The ATLAS tile calorimeter. Technical report ATL-TILECAL- PROC-2015-002. CERN, Geneva, Mar 2015. https://doi.org/10.1109/ANIMMA.2015.7465554. https://cds.cern.ch/record/2004868
21. Artamonov A et al (2008) The ATLAS forward calorimeter. J Instrum 3.02: P02010–P02010. https://doi.org/10.1088/1748-0221/3/02/p02010. https://doi.org/10.1088%2F1748-0221%2F3%2F02%2Fp02010
22. Gillberg D (2011) Performance of the ATLAS forward calorimeters in first LHC data. J Phys: Conf Ser 293:012041. https://doi.org/10.1088/1742-6596/293/1/012041. https://doi.org/10.1088%2F1742-6596%2F293%2F1%2F012041
23. ATLAS Collaboration (2010) ATLAS technical paper list of figures. https://twiki.cern.ch/twiki/bin/view/AtlasPublic/AtlasTechnicalPaperListOfFigures
24. ATLAS muon spectrometer (1997) Technical design report. Technical design report ATLAS. CERN, Geneva. https://cds.cern.ch/record/331068
25. Diehl E (2011) Calibration and Performance of the ATLAS Muon Spectrometer. Technical report. Comments: Proceedings of the DPF-2011 Conference, 7 pages, 15 figures. CERN, Geneva, Oct 2011. https://cds.cern.ch/record/1385884. arXiv:1109.6933
26. Pequenao J (2008) Computer generated image of the ATLAS Muons subsystem, Mar 2008. https://cds.cern.ch/record/1095929
27. Bauer F et al (2001) Construction and test of MDT chambers for the ATLAS Muon Spectrometer. Nucl Instrum Meth A461:17–20. https://doi.org/10.1016/S0168-9002(00)01156-6. arXiv:1604.02000 [physics.ins-det]
28. Argyropoulos Theodoros et al (2009) Cathode strip chambers in ATLAS: installation, commissioning and in situ performance. IEEE Trans Nucl Sci 56:1568–1574. https://doi.org/10.1109/TNS.2009.2020861
29. Aielli G et al (2006) The RPC first level muon trigger in the barrel of the ATLAS experiment. Nucl Phys Proc Suppl 158:11–15. [11(2006)]. https://doi.org/10.1016/j.nuclphysbps.2006.07.031
30. Majewski S et al (1983) A thin multiwire chamber operating in the high operation mode. Nucl Instrum Meth 217:265–271. https://doi.org/10.1016/0167-5087(83)90146-1
31. Nedden MZ (2017) The LHC Run 2 ATLAS trigger system: design, performance and plans. J Instrum 12.03:C03024-C03024. https://doi.org/10.1088/1748-0221/12/03/c03024. https://doi.org/10.1088%2F1748-0221%2F12%2F03%2Fc03024

# Chapter 5
# Physics Object Reconstruction

The particles produced in the *pp* collisions travel throughout the ATLAS detector volume, leaving different signatures in the sub-systems, according to the particle properties and the nature of the physics processes taking place. Figure 5.1 illustrates how different types of particles interact with the components of the ATLAS detector. The collected raw information are combined using dedicated algorithms, to reconstruct the relevant physics objects and identify the particles emerging from the beam collision. Before being used in physics analyses, the objects are corrected for detector effects, in a process called *calibration*.

In this chapter, a description of the reconstruction algorithms and techniques used to identify the objects detected at the ATLAS experiment is provided.

## 5.1 Tracks and Primary Vertices

The presence of an axial magnetic field along the $z$ direction forces the charged particles stemming from the hard collision to follow helical trajectories. The trajectory and momentum of charged particles are reconstructed by starting from the deposits of energy left by the particle in the ATLAS ID [1]. The energy deposits are clustered together to form *hits*. The main track recognition is based on an *inside-out* strategy [2], which starts with the identification of triplets of hits on different layers of the sub-detector, either Pixels or SCT modules. A combinatorial Kalman filter [3] is then used to extend the track by adding the most likely successive hits, according to the preliminary trajectory. This procedure results in a number of ambiguous track candidates, due to an incorrect assignment of the hits or overlapping tracks with shared hits. The following stage attempts to solve the ambiguities, by assigning a score to each candidate, based on the number of hits involved, the origin of each hit (the most

 
C. Tosciri, *Higgs Boson Decays into a Pair of Bottom Quarks*, Springer Theses,
https://doi.org/10.1007/978-3-030-87938-9_5

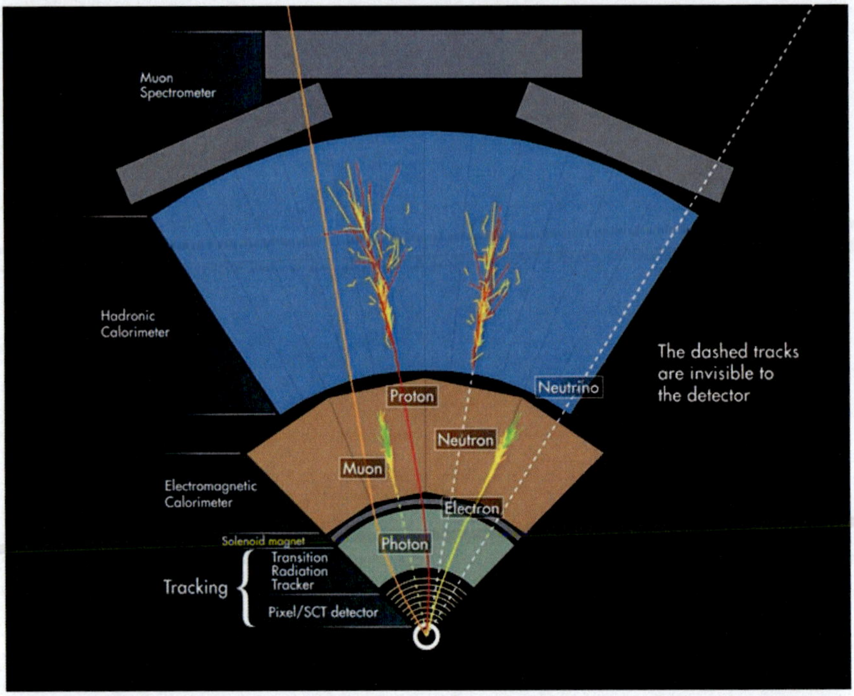


**Fig. 5.1** Transversal view of the particle paths in the ATLAS detector. Particles of different nature interact in different ways and in different sectors of the detector. Image *source* [4]

precise measurements coming from Pixels have a higher score) and the presence of *holes*, which are expected hits missing in the prediction trajectory. Additionally, a $\chi^2$ fit is performed on the tracks and poor-quality candidates are penalised. The momentum of the track is considered as well to promote energetic tracks and suppress low-$p_{\mathrm{T}}$ tracks, which are usually associated with wrong assignments. Poorly ranked candidates are removed, while the other tracks are extended to the TRT volume and combined with its measurements. In order to reconstruct those tracks which may have missed some hits in the silicon tracker, a reverse approach, called *outside-in*, is also applied to start a pattern recognition from the TRT module. Finally, all the reconstructed tracks are required to pass general criteria, like having a minimum of seven hits in the silicon detector, a maximum of two shared hits and a transverse momentum $p_{\mathrm{T}} > 400$ MeV. Tracks can be only reconstructed within the ATLAS ID pseudorapidity range of $\eta < 2.5$.

The precise reconstruction of the *primary vertex* [5, 6], the position where the hard scattering between the protons has occurred, is fundamental for the physics analyses. In a single triggered event, due to the harsh environment of pile-up, several interaction vertices can be reconstructed. The identification of the primary vertex involves an iterative procedure of finding and fitting the vertices. The reconstructed

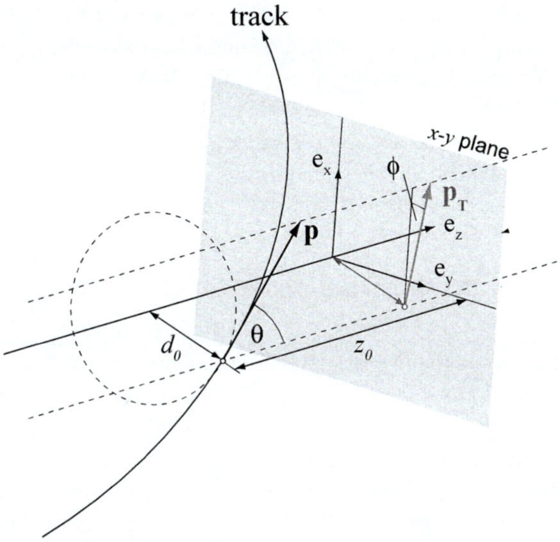


**Fig. 5.2** Transverse $d_0$ and longitudinal $z_0$ impact parameters of a track with respect to the interaction point. Image *source* [7]

tracks are used as input and combined with a seed position, selected for one vertex at a time, to determine the optimal vertex position with a $\chi^2$ minimisation. The tracks are weighted, according to their compatibility with the vertex candidate and then a new vertex position is computed. The procedure is iterated until a stopping criterion is reached and the vertex position determined. The tracks found incompatible with the current vertex are rejected and used to identify the next vertex, and the procedure is repeated until all the reconstructed tracks are associated with a corresponding vertex. Out of all the reconstructed vertices, the primary vertex is identified as the one having the highest sum of the squared transverse momenta of associated tracks: $\sum p_{\mathrm{T}}^2$. The remaining vertices are labelled as *pile-up vertices*.

In the reconstruction of the primary vertex, two important parameters are used: $d_0$ and $z_0$, respectively the transverse and longitudinal impact parameters, defined as the projections on the transversal and longitudinal planes of the distance between the interaction point and the *perigee*, the point of the track's closest approach to the beam axis. Impact parameters are illustrated in Fig. 5.2. The corresponding uncertainties are denoted as $\sigma_{d_0}$ and $\sigma_{z_0}$ and are estimated in the track fit. The efficiency of the primary vertex reconstruction is estimated to be higher than 99% [5].

## 5.2 Electrons

Electrons and positrons are reconstructed in the central region of the ATLAS detector ($|\eta| < 2.5$) by combining tracks reconstructed in the ID and energy deposits in the EM calorimeter, appearing as EM showers [8, 9]. Since the forward region with pseudorapidity $2.5 < \eta < 4.9$ is not covered by tracking detectors, only the information collected in the EMEC and FCal calorimeters can be exploited to reconstruct electrons, selected with $E_{\mathrm{T}} > 20$ GeV. To compensate for the absence of tracking information, variables describing the shape of the EM showers are employed. In the rest of this section, the reconstruction, identification and isolation will be discussed exclusively for central electrons.

### 5.2.1 *Reconstruction*

Electromagnetic *cluster seeds* are identified with a *sliding-window* algorithm which scans the $\eta \times \phi$ space of the EM calorimeter with a window of $3 \times 5$ units of size $\Delta\eta \times \Delta\phi = 0.025 \times 0.025$, corresponding to the granularity of the EM calorimeter middle layer. The energy found in the cells in all the longitudinal layers of the calorimeter within the window are summed into a *tower* energy and, in case this energy is larger than 2.5 GeV, a cluster seed is identified. Electromagnetic clusters are then formed around the seeds with the use of clustering algorithms [10]. The reconstructed tracks characterised by energy loss values which are compatible with an electron hypothesis (up to 30% energy loss at each intersection with the detector material) are combined with the EM clusters. In particular, a loose matching is performed between the extrapolated position of the track in the middle layer of the EM calorimeter and the barycentre of a cluster, if their angular distance satisfies $\Delta\eta < 0.05$ and $\Delta\phi < 0.1$. To take into account the significant amount of energy losses due to bremsstrahlung of high-energy electrons, a fitting algorithm, called *Gaussian Sum Filter* [11], refits the tracks that have enough precision hits and are loosely associated to a cluster. The matching is performed again with the refitted tracks and more stringent conditions. To reconstruct an electron candidate, at least one track is required to be matched to a seed cluster. If more than one track is matched to a cluster, tracks are ranked according to the number of precise hits and distance to the cluster $\Delta R = \sqrt{(\Delta\eta)^2 + (\Delta\phi)^2}$, to identify the primary track. Electron candidates with no associated precision hit tracks are removed and considered to be photons. The tracks associated to an electron candidate need to be compatible with the primary vertex of the hard scattering and satisfy the requirements on the impact parameters: $d_0/\sigma_{d_0} < 5$ and $\Delta z_0 sin\theta < 0.5$ mm, where $d_0$ and $z_0$ are the impact parameters of the track introduced in Sect. 5.1 and $\theta$ is the polar angle of the track.

### 5.2.2 *Identification*

In order to reduce the background arising from the conversion of photons into $e^+e^-$ pairs or due to secondary particles, for instance, coming from semileptonic decays of heavy-flavour hadrons, prompt electrons in the central region of the detector without transition regions, covering $\eta < 2.47$, are selected using electron identification algorithms [12]. In particular, a likelihood-based algorithm (LH) is applied. Track measurements data, such as number of hits in each layer, impact parameters and momentum lost by the track in traversing the ID, and electron cluster information, including shower width, energy and measurements of the bremsstrahlung effects, are combined by LH into an MVA. For a given electron, the signal and background probabilities are evaluated by combining these information and then a discriminant $d_{\mathcal{L}}$ is built as

$$d_{\mathcal{L}} = \frac{\mathcal{L}_S}{\mathcal{L}_S + \mathcal{L}_B} \quad \text{with} \quad \mathcal{L}_{S(B)}(\mathbf{x}) = \prod_{i=1}^{n} P_{S(B),i}(x_i), \tag{5.1}$$

where $\mathbf{x}$ is the vector containing all the mentioned discriminating variables and $P_{S(B),i}(x_i)$ is the probability density function of the variable $x_i$, in case of signal ($S$) or background ($B$) hypothesis, based on simulated events.

Fixed values of the LH discriminant are used to define different operating points, denominated as *VeryLoose*, *Loose*, *Medium*, and *Tight*. The operating points correspond to increasing thresholds for the LH discriminant and account for slightly different selections, which are more selective for *medium* and *tight* electrons, requiring, for instance, that at least one hit of the track must be in the innermost pixel layer. The electron candidates that satisfy tighter operating points also satisfy less restrictive operating points, in such a way that if an electron candidate is selected with *tight* criteria, it is certainly selected by the other operating points. The electron identification allows for an increasing background rejection as the selection gets tighter, nonetheless with a progressive cost for the efficiency of the selection itself. The minimum $E_{\mathrm{T}}$ of the electron identification was set to 4.5 GeV in Run-2.

### 5.2.3 *Isolation*

To further improve the background rejection, electrons are required to be isolated from additional tracks or energy deposits. A small activity in the $\Delta\eta \times \Delta\phi$ area surrounding the electron is a typical signature for prompt electrons. Several variables can be constructed using the calorimeter and tracking measurements, in order to quantify the amount of this activity. Calorimeter-based isolation criteria require to construct a cone of size $\Delta R$ around the candidate electron's cluster position and compute the isolation variable by summing up the energies deposited within the cone, excluding the energy deposited by the electron itself. Track-based isolation variables are constructed from the sum of the transverse momenta of the tracks with $p_{\mathrm{T}} > 1$ GeV found within a cone of radius $\Delta R$ constructed around the candidate's trajectory, excluding the electron track's own contribution. To minimise pile-up contamination, only tracks from the primary vertex are used in the definition of the isolation variable. Different criteria applied to the isolation variables give rise to different operating points for electron isolation. The choice depends on the specific needs of the physics analysis considered. The most relevant operating points for the analyses discussed in Sects. 7 and 8 are the following:

- *LooseTrack isolation*: targets a fixed value of 99% for the isolation efficiency by varying the radius of the isolation cone depending on the transverse momentum of the electron (track-based).
- *HighPtCalo isolation*: uses calorimeter-based isolation in a cone of fixed radius $\Delta R = 0.2$ and requires that the energy deposited does not exceed 3.5 GeV.

The electron efficiencies are evaluated both in data and in simulation from $Z \rightarrow e^+e^-$ and $J/\psi \rightarrow e^+e^-$ samples, for each of the steps previously described. The overall efficiency, obtained through the product of the reconstruction, identification and isolation efficiencies, is critical for physics measurements. The efficiencies obtained with data measurements are compared with simulations to retrieve correction *scale factors* as a function of $E_{\mathrm{T}}$ and $\eta$ of the electron candidate to account for MC mis-modelling. The scale factors obtained are very close to unity.

## 5.3 Photons

Similarly to electrons, energetic photons originate electromagnetic showers within the EM calorimeter. Prompt photons are reconstructed in the central region of the detector $|\eta| < 2.5$, following a similar strategy to the one used for prompt electrons [13].

### *5.3.1 Reconstruction and Identification*

Due to very similar signatures, the reconstruction of photons and electrons proceeds in parallel. The sliding-window algorithm introduced in Sect. 5.2 is used to identify seed clusters which are loosely matched to reconstructed ID tracks. Both photons that convert and do not convert to $e^+e^-$ pairs can be reconstructed, and are respectively referred to as *converted* and *unconverted* photons. The converted photons usually leave a *conversion vertex* which can be reconstructed from pairs of oppositely charged tracks in the ID that are likely to be electrons. The final discrimination between unconverted photons, converted photons and electrons is achieved through a set of decisions which take into account the presence or absence of hits in the silicon detector and conversion vertices. MC simulations show that the 96% of prompt photons with $E_{\mathrm{T}} > 25$ GeV are expected to be reconstructed as photon candidates, while the remaining 4% are incorrectly reconstructed as electrons.

Prompt photons need to be identified against background photons, usually arising from hadronic jets. The identification is based on cuts performed on a set of variables which describe the shape and other characteristics of the showers. Electromagnetic showers originated from prompt photons are typically narrower than the ones from hadronic jets and are associated with a smaller leakage in the HCAL, indicating that no additional hadrons are present. Photons arising from the decay of isolated neutral pions ($\pi^0 \rightarrow \gamma\gamma$) are often characterised by two separate local energy maxima in the first layer of the ID, due to the small separation between the two photons. The separation between prompt and background photons is complicated by the presence of low-$E_T$ activity in the detector, due to pile-up events. Two reference operating points, *Loose* and *Tight*, are defined according to the selection criteria.

## 5.4 Muons

Muons ($\mu^{\pm}$) are reconstructed within the acceptance of the muon spectrometer with $\eta < 2.7$. After leaving hits in the ATLAS ID, these particles deposit a small amount of energy in the calorimeters and then reach the ATLAS MS [14].

### *5.4.1 Reconstruction*

Muon reconstruction is performed separately in each sub-system and then combined to construct muon track candidates. The track reconstruction in the ID follows the general strategy described in Sect. 5.1. In the MS, muon tracks are built from segments reconstructed using hits found within each set of muon chambers. The segments are then combined through a fit performed on the hits, according to criteria based on relative positions and angles of the hits and multiplicity. The segment combination forms the muon track.

The combination of ID and MS tracks, with the additional use of calorimeter information, can be done with various algorithms which define different reconstructed muon types, as listed below:

- *Combined muons* (CB): after independent track reconstruction in the ID and MS, a combined track is built by fitting the hits from the two sub-detectors. Muons are mostly reconstructed following an outside-in pattern recognition which starts with hits collected in the MS and then extrapolates the track into the ID. This is the main type of reconstructed muons and provides the highest muon purity.
- *Segment-tagged muons* (ST): a track in the ID is associated with at least one track segment in the MS. Mostly based on the ID information, this type is used when muons cross only one layer of the MS, due to low $p_{\mathrm{T}}$ or because they fall outside the MS acceptance.
- *Calorimeter-tagged muons* (CT): a track in the ID is combined with an energy deposit in the calorimeter. This type is used when track information from the MS is missing, due to not instrumented regions.

- *Extrapolated muons* (ME): a track is reconstructed exclusively from MS segments, requiring information at least from two layers of MS chambers. In the forward region, muons are reconstructed at least from three layers of the MS. ME muons are mainly used to extend the acceptance in the region of pseudorapidity $2.5 < |\eta| < 2.7$, which is not covered by the ID.

### 5.4.2 Identification

After reconstruction, prompt muons from signal processes must be identified against the background, mainly arising from in-flight decays of *pions* $\pi$ and *kaons* $K$. A typical characteristic of background muons is a 'kink' shown in the reconstructed track. Several variables can be constructed to impose quality requirements that suppress this kind of background. For instance, the *q/p significance*[1] quantifies the incompatibility between the momentum measured in the ID and the one measured in the MS, while a certain value of the $\chi^2$ of the combined fit reveals the poor quality of the fit in case of background. Different muon operating points are provided, according to the tightness of the required selection criteria: *Loose*, *Medium*, *Tight* and *High-$p_T$*. The *Loose* selection offers the best efficiency (98.1%) with slightly worst purity while, as the selection gets tighter, the purity improves with a cost on the efficiency (in the case of *Tight* selection, the efficiency is up to 89.9%).

### 5.4.3 Isolation

In order to disentangle prompt muons decaying from bosons such as $W$, $Z$ and $H$, and muons arising from the semileptonic decays of heavy-flavour hadrons, muon isolation criteria are applied on specific isolation variables, constructed using tracks and calorimeter information. The track-based isolation variable $p_{\mathrm{T}}^{varcone30}$ is the scalar sum of the tracks transverse momenta in a $p_{\mathrm{T}}$-dependent variable cone with $\Delta R = min(10\,\mathrm{GeV}/p_{\mathrm{T}}^{\mu}, 0.3)$, where $p_{\mathrm{T}}^{\mu}$ is the muon transverse momentum built around the muon track. $E_{\mathrm{T}}^{cone20}$ is a calorimeter-based isolation variable, constructed as the sum of the transverse energy of calorimeter clusters in a cone of fixed size $\Delta R = 0.2$ built around the muon. Several working points are defined for the reconstructed muons, according to the isolation cut. The relevant working points for the analyses discussed in Sects. 7 and 8 are the following:

- *LooseTrack*: requires a variable cut on the variable $p_{\mathrm{T}}^{varcone30}/p_{\mathrm{T}}$, where $p_{\mathrm{T}}$ is the transverse momentum of the muon, in such a way to obtain 99% of efficiency in the whole $\eta$ and $p_{\mathrm{T}}$ range.

[1] *q/p significance* is defined as $\sigma(q/p) = \frac{|(q/p)^{ID} - (q/p)^{MS}|}{\sqrt{(\sigma(q/p)^{ID})^2 + (\sigma(q/p)^{MS})^2}}$, where $q$ and $p$ are the charge and momentum of the muon.

- *HighPtTrack*: requires $p_{\mathrm{T}}^{cone20}/p_{\mathrm{T}} < 1.25$ GeV and provides up to 95% of efficiency.

The efficiencies for the isolation working points are measured in data and simulation in $Z \rightarrow \mu^+\mu^-$ and $J/\psi \rightarrow \mu^+\mu^-$ decays. Similarly to the electron reconstruction, the isolation scale factors are evaluated to correct for simulation mis-modelling.

## 5.5 Jets

The ultimate goal of physics analyses is to reconstruct the fundamental particles produced in the primary interaction. In a high-energy $pp$ collision at the LHC the most frequently produced particles are coloured quarks and gluons. However, as described in Sect. 2, quarks and gluons can not be observed as free particles because, after their production, they immediately undergo *fragmentation* and *hadronisation*, processes responsible for the generation of energetic colourless hadrons. These particles, collimated in streams called *jets*, deposit a large amount of energy in the ATLAS calorimeters. A jet is the most complicated object to be reconstructed in ATLAS and, ideally, measuring its energy and direction provides information about the original parton that initiated it [15].

### 5.5.1 Reconstruction

Additionally to the sliding-window algorithm introduced in Sect. 5.2.1, calorimeter clusters are reconstructed in ATLAS by merging topologically-related energy deposits of the electromagnetic and hadronic calorimeters, following a signal-significance pattern recognition [16]. The resulting *topo-clusters* are used to reconstruct jets through a clustering algorithm that repeatedly recombines the closest pair of clusters, according to some distance measures. This algorithm, implemented in the `FastJet` [17] package, is known as the *anti-$k_T$ algorithm* [18] and relies on the following distance measures:

$$d_{ij} = \min(k_{ti}^{-2}, k_{tj}^{-2})\frac{\Delta R_{ij}^2}{R^2}, \qquad d_{iB} = k_{ti}^{-2}, \tag{5.2}$$

where $\Delta R_{ij}^2 = (y_i - y_j)^2 + (\phi_i - \phi_j)^2$ is the angular distance between two objects $i$ and $j$, and $k_{ti}$, $y_i$ and $\phi_i$ are respectively the transverse momentum, rapidity and azimuth of the object $i$. The clustering procedure compares $d_{ij}$, which represents the distance between two objects $i$ and $j$, with $d_{iB}$, the distance of the object $i$ with respect to the beam. If $d_{ij} < d_{iB}$, the objects $i$ and $j$ are recombined together and the algorithm proceeds with the next iteration. If $d_{ij} > d_{iB}$, the object $i$ is reconstructed as a jet and removed from the list of objects. The distances are recalculated and the

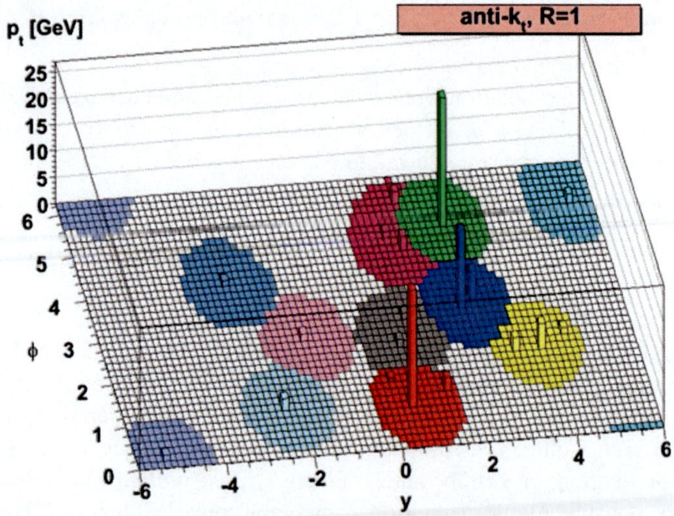


**Fig. 5.3** Simulation of jet reconstruction as results with the *anti-$k_T$* algorithm. Image *source* [18]

procedure repeated until no objects are left. The *radius parameter* $R$ defines the size of the maximum jet cone. The algorithm works in such a way that if a hard object has no hard neighbours within a distance of $2R$, it just accumulates all the soft objects within a circle of radius $R$, with the consequence that the hard jets are all circular with a radius $R$, as illustrated in Fig. 5.3. Values of $R = 0.4$ and $R = 1.0$ are usually considered for the jet radius parameter. The first value is typically used for jets initiated by gluons or quarks, also called *small jets*. The other choice is used for jets containing the hadronic decays of massive particles, such as $W$, $Z$ and Higgs bosons or top quarks, and they are usually referred to as *fat jets*.

### 5.5.2 Calibration

The energy of the reconstructed jets has to be calibrated to account for several effects, such as the calorimeter non-compensation (regarding the different response for electromagnetic and hadronic calorimeters), dead material and leakage, *out-of-cone* energy (by construction, not included in the jet) and pile-up contamination. Through calibration, the jet energy scale (JES) of the reconstructed jet can be restored to the scale of truth jets. The calibration is derived using a combination of methods based on MC simulation and data-driven techniques, summarised in the following [19].

**Origin correction** To improve the $\eta$ resolution, the direction of the jet is corrected to make it point to the hard-scatter primary vertex rather than the centre of the detector, while keeping constant the jet energy.

**Pile-up corrections** Pile-up interactions can occur within the bunch crossing of interest (*in-time*) or in neighbouring bunch crossings (*out-of-time*). The corrections aim at removing both of the two contributions and proceed in two steps: first, the average contribution is subtracted by the $p_{\mathrm{T}}$ of the jet with a method based on the area of the jet; then, residual pile-up dependencies are removed, as a function of $\mu$ (the number of interactions per crossing, defined in Sect. 4.1.2) and the number of reconstructed primary vertices.

**Absolute JES correction** It is performed with a MC simulation and is based on the comparison between reconstructed jets, calibrated up to the pile-up corrections, and truth jets, selected with a geometrical matching with $\Delta R = 0.3$. The average energy response is derived from the mean value of the Gaussian fit of $E^{\mathrm{reco}}/E^{\mathrm{truth}}$ distribution, and is obtained as a function of $E^{\mathrm{truth}}$ and $\eta$. A numerical inversion procedure provides the corrections in $E^{\mathrm{reco}}$ from $E^{\mathrm{truth}}$. An additional correction is derived as the difference between the reconstructed $\eta^{\mathrm{reco}}$ and $\eta^{\mathrm{truth}}$, parametrised as a function of $E^{truth}$ and $\eta$, to account for the transition regions between the barrel and end-cap ($\eta \sim 1.4$) and between the end-cap and forward ($\eta \sim 3.1$) calorimeters.

**Eta inter-calibration** Well-measured jets in the central region of the detector ($\eta <$ 1.4) are used to derive a residual calibration for jets in the forward region ($0.8 < |\eta| < 4.5$).

**Global Sequential Calibration (GSC)** Aiming to reduce fluctuations of the jet response due to energy leakage, five observables are identified and for each of them an independent jet four-momentum correction is derived as a function of $p_{\mathrm{T}}^{\mathrm{truth}}$ and $|\eta|$. Correcting the selected variables, which depend on the shape and energy of the jet, improves the distinction between quark- and gluon-initiated jets.

***In-situ* calibration** As the last stage of the jet calibration, it corrects for differences between MC and data, arising from an incorrect detector description in the MC. The JES is calibrated in data, using as reference a well-calibrated object. A method called *direct balance* is used which exploits $Z + jet$ processes, in case of low-$p_{\mathrm{T}}$ region [20–500] GeV, and $\gamma + jet$ processes in the middle-$p_{\mathrm{T}}$ region [36–950] GeV. The $p_{\mathrm{T}}$ balance between the reference boson and the jet, $p_{\mathrm{T}}^{ref}/p_{\mathrm{T}}^{jet}$, is fitted with a Gaussian fit to extract the corrections. The *multijet balance* technique extends the calibration up to $p_{\mathrm{T}} = 2$ TeV and, rather than balancing the reference object against the jet, it uses a whole hadronic recoil.

After applying the set of corrections, the JES is evaluated as a function of the jet transverse momentum. The jet energy resolution (JER) defined as $\sigma_{p_{\mathrm{T}}}/p_{\mathrm{T}}$ is also obtained in-situ by comparing the width of the balance distribution in data and MC [20]. The results are shown in Fig. 5.4. The JES and JER corrections are evaluated together with a large set of systematic uncertainties, mainly arising from the data-driven stages and accounting for assumptions made on the MC simulation, sample statistics, propagation of the uncertainties of the energy scales of other objects. MC

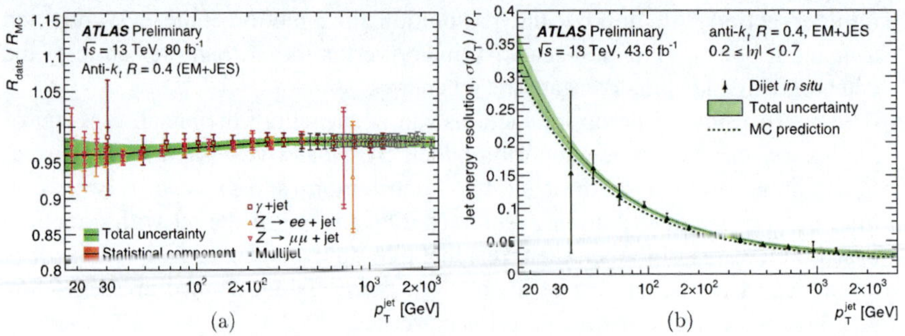


**Fig. 5.4** Data-to-MC ratio of the average jet $p_T$ response as a function of jet $p_T$. The combined result is based on three in-situ techniques: the $Z$+jet balance method, $\gamma$+jet balance method and the multijet balance (**a**). The relative JER $\sigma_{p_T}/p_T$ as a function of $p_T$. The results are obtained for anti-kt jets with a radius parameter of $R = 0.4$, calibrated following the JES scheme, including the residual in-situ calibration and using the 2017 dataset (**b**). Image *sources* [21, 22]

mis-modelling of pile-up parameters also affect the accuracy of the corrections, as well as differences between MC and data for the jet flavour composition. Systematic uncertainties are determined by varying the parameters of the object selection, in addition to ascertaining the MC effect with alternative MC samples, and comparing the *in-situ* JER with the resolution obtained on MC only (Closure).

### *5.5.3 Pile-Up Jets Suppression*

Pile-up suppression is fundamental for most physics analyses. After the dedicated correction introduced in the previous section, spurious jets may remain, due to local fluctuations in the pile-up activity. The *jet-vertex fraction* (JVF) variable measures the fraction of transverse momentum from the tracks that are associated with the hard-scatter vertex. In Run-1 JVF was used to separate jets originated from the hard-scatter and those from pile-up interactions by imposing a minimum threshold on it. However, this procedure leads to a dependency of the jet efficiency on the number of reconstructed primary vertices in the event, $N_{vtx}$. New variables, which are based on tracking information and are stable as a function of $N_{vtx}$, have been defined and a multivariate combination of them, called the *jet-vertex-tagger* (JVT) algorithm, has been developed for Run-2. The discriminant is constructed using the variables as a 2-dimensional likelihood [23]. Its output is shown in Fig. 5.5.

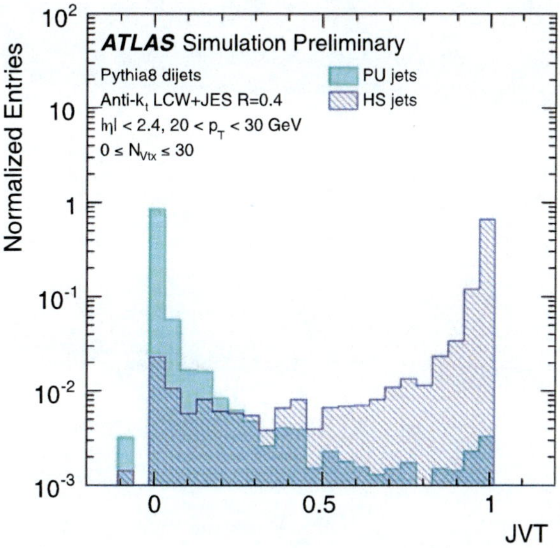


**Fig. 5.5** Distribution of the JVT discriminator for pile-up and hard-scatter jets. Image *source* [23]

### 5.5.4 b-*Jets Tagging*

The identification of the flavour of the quark originating the jets is of fundamental importance for every physics analysis, to properly reconstruct the primary process happened in the hard-scattering. In particular, flavour identification, or *flavour tagging*, is vital for many analyses with $b$-quark jets in the final state, such as the $VH(H \to b\bar{b})$ analyses, discussed in Sects. 7 and 8. Flavour tagging is enabled by the distinctive properties of the heavy hadrons produced in the hadronisation of the original quarks. A $b$-jet is originated from the hadronisation of a $b$-quark at the primary vertex and the consequent decay of the newly formed $b$-hadron. The main property of a $b$-hadron is the relatively long lifetime of order $1.5 \times 10^{-12}$ s, which results in a measurable displacement of the decay point, the secondary vertex, with respect to the primary vertex. The secondary vertex is sketched in Fig. 5.6. Depending on the momentum, the secondary vertex is typically a few millimetres away from the primary vertex. Additionally, because of the large mass of $\sim 5$ GeV, the $b$-hadron retains a large fraction of the $b$-quark momentum ($\sim 75\%$) and decays into a large number of decay products, resulting in a large track multiplicity. Because of geometrical effects, the tracks arising from the secondary vertex tend to have a large impact parameter with respect to the primary vertex. Furthermore, a $b$-quark decays preferably into a $c$-quark, due to the CKM transition matrix values $|V_{cb}|^2 > |V_{ub}|^2$ [25]. Therefore, $b$-jets very often contains $c$-hadrons which in turn decay with their own detectable displaced vertex [26].

The flavour tagging techniques applied in ATLAS are designed to exploit $b$-jet characteristics, in order to discriminate between $b$-jets, $c$-jets and jets originating from light quarks ($u$, $d$, $s$), called $light$-jets [27]. The ATLAS strategy is based on

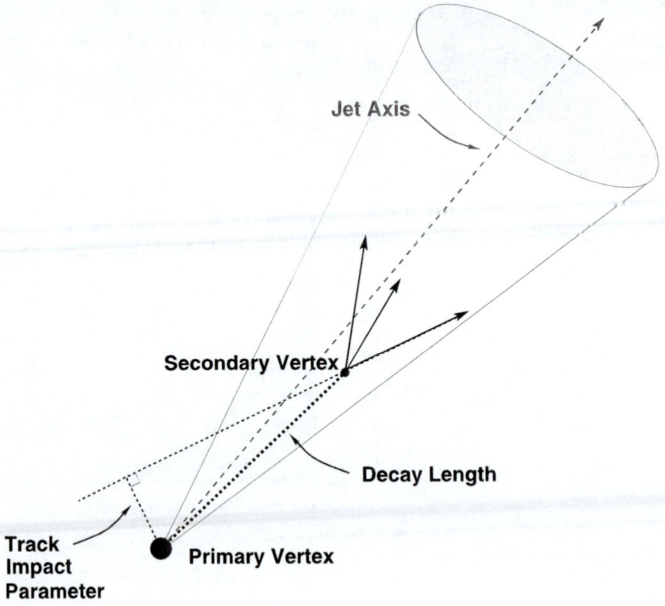


**Fig. 5.6** Illustration of the principle of $b$-hadron decay. Image *source* [24]

two different levels of flavour taggers. Low-level taggers use a series of algorithms that aim at extracting the features that are relevant for the discrimination of the flavour nature of the jets. The outputs of these algorithms are fed into a high-level tagger, whose job is to estimate a total likelihood ratio for the $b$-jet versus no-$b$-jet hypothesis. In the following sections, the main $b$-tagging algorithms used in the analyses presented in Sects. 7 and 8 are briefly described.

### Impact Parameter Based Algorithms

Due to the long lifetime of $b$-hadrons, tracks generated from $b$-hadron decay products show impact parameters larger than tracks emerging from the primary vertex. This property is exploited by impact parameter based taggers [28]. The sign of the impact parameter of a track with respect to the primary vertex can be assigned by determining whether the point of closest approach of the track to the primary vertex is in front (positive) or behind (negative) the primary vertex with respect to the jet direction. Tracks from $b$- and $c$-hadrons tend to have a positive sign, while background is usually related to a negative sign. Figure 5.7 shows the distribution of the track signed impact parameter significance with respect to the primary vertex, in the transversal ($d_0/\sigma_{d_0}$) and longitudinal ($z_0 \sin\theta/\sigma_{z_0}$) planes. The probability density functions for the signed impact parameter significance of these tracks are used to define ratios of

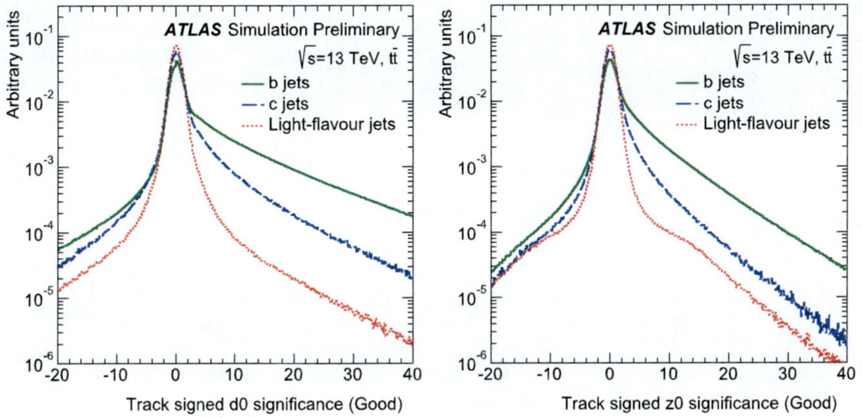


**Fig. 5.7** The transverse (**a**) and longitudinal (**b**) signed impact parameter significance of tracks in events associated with $b$ (green), $c$ (blue) and $light$-flavour (red) jets. Image *sources* [28]

the $b$- and $light$-flavour jet hypotheses. The final discriminant is then derived using a log-likelihood ratio obtained by the combination of the probability density functions.

### Secondary Vertex Finding Algorithm

The Secondary Vertex algorithm (SV) [29] attempts to reconstruct the secondary vertex of the $b$-hadron decay, while rejecting vertices likely to originate from the decay of long-lived particles (like $K_S$ or $\Lambda$), photon conversion or hadronic interactions with the detector material. It also constructs a set of observables from specific properties related to the reconstructed secondary vertex, which are used as input to the high-level discriminator. Some of these variables are shown in Fig. 5.8: the invariant mass of the tracks associated to the vertex $m$(SV), which is indicative to the mass of the hadron that initiates the jet; the transverse decay length $L_{xy}$; the distance between the primary and secondary vertices divided by its uncertainty $S_{xyz}$; the energy fraction $f_E$, defined as the energy from the tracks in the displaced vertex relative to all the tracks reconstructed within the jet.

### Decay Chain Multi-vertex Algorithm

The decay chain multi-vertex reconstruction algorithm, called `JetFitter`, is dedicated to the reconstruction of the full decay chain from the primary to the secondary vertex of the $b$-hadron decay and finally to the tertiary vertex, due to the further decay of a $c$-hadron. The algorithm uses a Kalman filter and assumes that the three vertices lie on a common line. With this approach, the $b$- and $c$-hadron vertices can

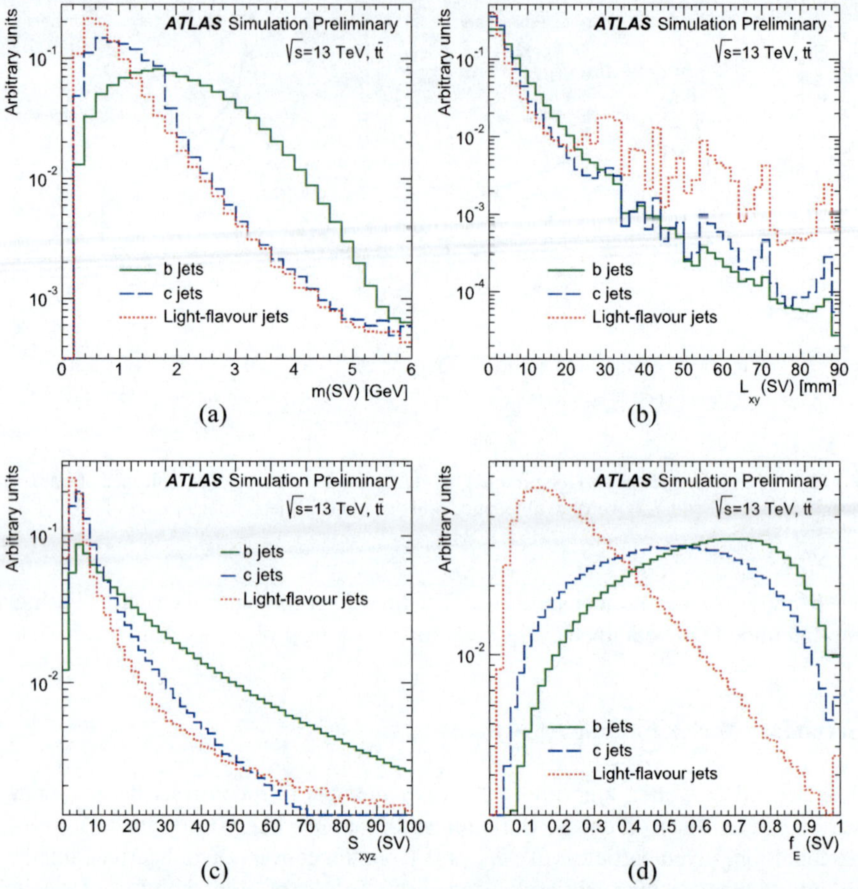


**Fig. 5.8** Properties of the secondary vertices reconstructed by the SV algorithm for $b$ (solid green), $c$ (dashed blue) and $light$-flavour (dotted red) jets: the invariant mass (**a**), the transverse decay length (**b**), the 3D decay length significance (**c**) and the energy from the tracks in the displaced vertex relative to all the tracks reconstructed within the jet (**d**). Image *source* [28]

be resolved [28]. A set of variables describing the properties of the decay topology and secondary vertices is reconstructed by `JetFitter`.

### Multivariate Algorithm

The outputs obtained with the low-level algorithms previously described, are fed into a high-level tagger, a multivariate algorithm which uses a BDT called MV2c10 [30]. This algorithm was developed for Run-2 and used in the $VH(H \to b\bar{b})$ analyses presented in Sects. 7 and 8. The transverse momentum and pseudorapidity of the jets are provided to the BDT as well, to take advantage of correlations with other

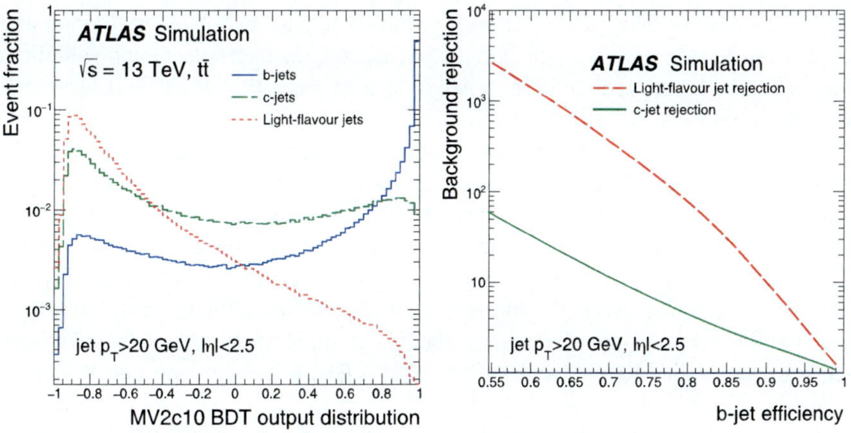


(a) (b)

**Fig. 5.9** The MV2c10 output for $b$-jets (solid line), $c$-jets (dashed line) and light-flavour jets (dotted line) in simulated $t\bar{t}$ events (**a**). The light-flavour jet (dashed line) and $c$-jet rejection factors (solid line) as a function of the $b$-jet tagging efficiency of the MV2c10 $b$-tagging algorithm (**b**). Image *sources* [30]

variables. The BDT training is performed on a subset of events from a simulated $t\bar{t}$ sample, by considering $b$-jets as signal and $c$-jets and light-flavour jets as background. The output of the MV2c10 algorithm is presented in Fig. 5.9, together with the background rejection factors for $c$-jets and light-flavour jets as a function of the $b$-jet efficiency. A set of operating points is defined according to selection criteria applied to the discriminant obtained by the $b$-tagging algorithm, designed to fix a certain efficiency in the simulated $t\bar{t}$ sample. The selection criteria can rely on a *single-cut* or a *flat efficiency* with variable cut on the BDT output. The single-cut operating point provides the targeted efficiency when averaged over the transverse momentum distribution of $b$-jets. The target efficiencies for the various operating points are 60%, 70%, 77% and 85%. Data-to-simulation scale factors are then calculated for each operating point as a function of the $b$-jet $p_{\mathrm{T}}$, by comparing the efficiency measured using collision data $\epsilon_b^{data}$ to that predicted by the simulation $\epsilon_b^{sim}$.

## 5.6 Tau Leptons

With a mass of 1.78 GeV [31], tau particles ($\tau^{\pm}$) are the heaviest among the lepton family. This is reflected in a very small lifetime of $(290.3 \pm 0.5) \times 10^{-15}$ s, corresponding to a distance $c\tau = 87\,\mu$m traveled by the $\tau$ leptons before decaying inside the beam pipe, preventing them to reach the active detectors. $\tau$ leptons decay either leptonically, with branching ratio of 35% or hadronically, in the remaining cases. The leptonic decays, $\tau \rightarrow \ell\nu_\ell\nu_\tau$, yield in the final state a lepton (electron or muon) which can not be distinguished from a prompt one and whose reconstruction and identification is very challenging. On the other hand, the hadronic component of tau decays can be typically identified in narrow jets of particles left in the calorimeter, associated with a number of tracks in the ID [32]. Therefore in ATLAS, tau reconstruction and identification concentrate on the hadronic decay modes, which typically lead to final states with one or three charged pions, called *prongs*, with the possible additional presence of one or two neutral pions which decay to photons ($\pi^0 \rightarrow \gamma\gamma$). Jets formed using the anti-$k_t$ algorithm with a distance parameter $R = 0.4$ are used in the tau reconstruction algorithm if $p_{\mathrm{T}} > 20$ GeV and $|\eta| < 2.5$. Additionally, the production vertex of the tau lepton is identified to increase the reconstruction efficiency and reduce pile-up effects, and either one or three tracks are associated with it. The tracks are required to be within the region $\Delta R < 0.2$ around the $\tau$-jet direction. Some requirements are imposed on the impact parameters of the tracks with respect to the tau vertex: $|d_0| < 1.0$ mm and $|\Delta z_0 sin\theta| < 1.5$ mm. An identification step is necessary to distinguish between $\tau$-jets and QCD jets, by exploiting the characteristics of the former, such as the low track multiplicity, collimated energy deposits, and the displacement of the tau vertex. Several discriminating variables are fed into a BDT, which is trained separately for 1-prong and 3-prong decays using samples of the process $Z/\gamma^* \rightarrow \tau\tau$. Three working points, labelled *Tight*, *Medium* and *Loose* and corresponding to different tau identification efficiency values, are defined.

## 5.7 Missing Transverse Momentum

For weakly interacting particles produced in the hard scattering, such as neutrinos or dark matter candidates, the probability of interacting within any of the ATLAS sub-detectors is negligible. By escaping detection, these particles cause an energy imbalance in the observed event. The conservation of momentum in the transversal plane of the detector requires the overall transverse momentum of the event to be zero. An imbalance can then be attributed to a weakly interacting particle produced in the hard collision and traversing the detector.

The imbalance can be measured through the reconstructed missing transverse momentum $\mathbf{E}_{\mathrm{T}}^{\mathrm{miss}}$, which is constructed by considering two contributions [33]. The first comes from the objects produced in the hard collision, comprising particles and jets fully reconstructed and calibrated. The second contribution is a *soft term* which accounts for additional charged-particle tracks that are associated with the primary vertex but not with any of the reconstructed hard objects. The missing transverse momentum vector is calculated as the vectorial sum of missing transverse momentum terms:

$$
\begin{aligned}
\mathbf{E}_{\mathrm{T}}^{\mathrm{miss}} &= -\left( \sum_{\text{electrons}} \mathbf{p}_T^{e} + \sum_{\text{photons}} \mathbf{p}_T^{\gamma} + \sum_{\tau\text{-leptons}} \mathbf{p}_T^{\tau_{\mathrm{had}}} + \sum_{\text{muons}} \mathbf{p}_T^{\mu} + \sum_{\text{jets}} \mathbf{p}_T^{\text{jet}} + \sum_{\text{tracks}} \mathbf{p}_T^{\text{track}} \right) \\
&= \mathbf{E}_{\mathrm{T}}^{\mathrm{miss,\,e}} + \mathbf{E}_{\mathrm{T}}^{\mathrm{miss,}} + \mathbf{E}_{\mathrm{T}}^{\mathrm{miss,}\,\emptyset_{\mathrm{had}}} + \mathbf{E}_{\mathrm{T}}^{\mathrm{miss,}\,^-} + \mathbf{E}_{\mathrm{T}}^{\mathrm{miss,\,jet}} + \mathbf{E}_{\mathrm{T}}^{\mathrm{miss,\,soft}}
\end{aligned}
\tag{5.3}
$$

Particles and jets included in the *hard term* are reconstructed with the methods described previously in this chapter, and are additionally accepted by the event selection. The object selection is refined in the context of every specific physics analysis, in order to optimise the $\mathbf{E}_{\mathrm{T}}^{\mathrm{miss}}$. Since the physics objects are reconstructed from signals collected within different sub-detectors, the same signal can be used to reconstruct different objects. To avoid this multiple counting, a hierarchy is defined for the physics object, prioritising electrons and then, in order, photons, tau leptons and jets. Muons do not show signal overlap, since they are reconstructed from MS tracks that are not reached by other objects. The soft term is reconstructed exclusively from ID tracks associated to the primary vertex and represents a very important contribution for the improvement of the $\mathbf{E}_{\mathrm{T}}^{\mathrm{miss}}$ scale and resolution, especially in final states with a low hard-objects multiplicity.

The reconstruction and measurement of the $\mathbf{E}_{\mathrm{T}}^{\mathrm{miss}}$ are very challenging tasks, due to the nature of this quantity. The uncertainties arising from the reconstruction of every single object contributing to the missing transverse momentum must be taken into account and combined into an overall systematic uncertainty associated with the measurement of $\mathbf{E}_{\mathrm{T}}^{\mathrm{miss}}$.

## References

1. Aaboud M et al (2017) Performance of the ATLAS track reconstruction algorithms in dense environments in LHC Run 2. Eur Phys J C 77(10):673. ISSN: 1434-6052. https://doi.org/10.1140/epjc/s10052-017-5225-7
2. Cornelissen T et al (2008) The new ATLAS track reconstruction (NEWT). J Phys: Conf Ser 119(3):032014. https://doi.org/10.1088/1742-6596/119/3/032014
3. Emil Kalman R (1960) A new approach to linear filtering and prediction problems. J Basic Eng 82(1): 35. https://doi.org/10.1115/1.3662552
4. Pequenao J, Schaffner P (2013) How ATLAS detects particles: diagram of particle paths in the detector. https://cds.cern.ch/record/1505342
5. Aaboud M et al (2017) Reconstruction of primary vertices at the ATLAS experiment in Run 1 proton-proton collisions at the LHC. Eur Phys J C 77(5). ISSN: 1434-6052. https://doi.org/10.1140/epjc/s10052-017-4887-5
6. Performance of primary vertex reconstruction in proton-proton collisions at $\sqrt{s} = 7TeV$ in the ATLAS experiment. Technical report ATLAS-CONF-2010-069. Geneva: CERN, July 2010. https://cds.cern.ch/record/1281344
7. Cornelissen TG et al (2007) Updates of the ATLAS tracking event data model (Release 13). Technical report ATL-SOFT-PUB-2007-003. ATL-COM-SOFT-2007-008. Geneva: CERN. https://cds.cern.ch/record/1038095
8. ATLAS Collaboration (2016) Electron efficiency measurements with the ATLAS detector using the 2015 LHC proton-proton collision data. Technical report ATLAS-CONF-2016-024. Geneva: CERN. https://cds.cern.ch/record/2157687
9. Aad G et al (2014) Electron reconstruction and identification efficiency measurements with the ATLAS detector using the 2011 LHC proton-proton collision data. Eur Phys J C CERN-PH-EP-2014-040. CERN-PH-EP-2014-040 74:38. https://doi.org/10.1140/epjc/s10052-014-2941-0. https://cds.cern.ch/record/1694142

10. Lampl W et al (2008) Calorimeter clustering algorithms: description and performance. Technical report ATL-LARG-PUB-2008-002. ATL-COM-LARG-2008-003. Geneva: CERN. http://cds.cern.ch/record/1099735
11. Improved electron reconstruction in ATLAS using the Gaussian Sum Filter-based model for bremsstrahlung. Technical report ATLAS-CONF-2012-047. Geneva: CERN (2012). https://cds.cern.ch/record/1449796
12. Aaboud M et al (2019) Electron reconstruction and identification in the ATLAS experiment using the 2015 and 2016 LHC proton-proton collision data at $\sqrt{s} = 13TeV$. Eur Phys J. C 79:639. 40 p. https://doi.org/10.1140/epjc/s10052-019-7140-6. arXiv:1902.04655
13. Aaboud M et al (2019) Measurement of the photon identification efficiencies with the ATLAS detector using LHC Run 2 data collected in 2015 and 2016. Eur Phys J C79(3):205. https://doi.org/10.1140/epjc/s10052-019-6650-6. arXiv: 1810.05087 [hep-ex]
14. Aad G et al (2016) Muon reconstruction performance of the ATLAS detector in proton- proton collision data at $\sqrt{s} = 13TeV$. Eur Phys J C 76(5):292. https://doi.org/10.1140/epjc/s10052-016-4120-y. arXiv: 1603.05598 [hep-ex]
15. Salam GP (2010) Towards jetography. Eur Phys J C 67:637–686. https://doi.org/10.1140/epjc/s10052-010-1314-6. arXiv: 0906.1833 [hep-ph]
16. Aad G et al (2017) Topological cell clustering in the ATLAS calorimeters and its performance in LHC Run 1. Eur Phys J C 77(7):490. ISSN: 1434-6052. https://doi.org/10.1140/epjc/s10052-017-5004-5
17. Cacciari M, Salam GP, Soyez G (2012) FastJet user manual. Eur Phys J C 72:1896. https://doi.org/10.1140/epjc/s10052-012-1896-2. arXiv: 1111.6097 [hep-ph]
18. Cacciari M, Salam GP, Soyez G (2008) The anti-ktjet clustering algorithm. J High Energy Phys 2008(04):063–063. https://doi.org/10.1088/1126-6708/2008/04/063
19. Aaboud M et al (2017) Jet energy scale measurements and their systematic uncertainties in proton-proton collisions at $\sqrt{s} = 13TeV$ with the ATLAS detector. Phys Rev D96(7):072002. https://doi.org/10.1103/PhysRevD.96.072002. arXiv: 1703.09665 [hep-ex]
20. Aad G et al (2013) Jet energy resolution in proton-proton collisions at $\sqrt{s} = 7TeV$ recorded in 2010 with the ATLAS detector. Eur Phys J C 73(3):2306. https://doi.org/10.1140/epjc/s10052-013-2306-0. arXiv: 1210.6210 [hep-ex]
21. ATLAS Collaboration (2018) Jet energy scale and uncertainties in 2015-2017 data and simulation. https://atlas.web.cern.ch/Atlas/GROUPS/PHYSICS/PLOTS/JETM-2018-006/
22. ATLAS Collaboration (2018) Jet energy resolution in 2017 data and simulation. https://atlas.web.cern.ch/Atlas/GROUPS/PHYSICS/PLOTS/JETM-2018-005/
23. Tagging and suppression of pileup jets with the ATLAS detector. Technical report ATLAS-CONF- 2014-018. Geneva: CERN (2014). https://cds.cern.ch/record/1700870
24. Performance of the ATLAS Secondary Vertex b-tagging Algorithm in 7 TeV Collision Data. Technical report ATLAS-CONF-2010-042. Geneva: CERN (2010). https://cds.cern.ch/record/1277682
25. Makoto K, Toshihide M (1973) CP violation in the renormalizable theory of weak interaction. Prog Theor Phys 49:652–657. https://doi.org/10.1143/PTP.49.652
26. Piacquadio G, Weiser C (2008) A new inclusive secondary vertex algorithm for b-jet tagging in ATLAS. 119(3):032032. https://doi.org/10.1088/1742-6596/119/3/032032
27. Optimisation of the ATLAS b-tagging performance for the 2016 LHC Run. Technical report ATL-PHYS-PUB-2016-012. Geneva: CERN (2016). https://cds.cern.ch/record/2160731
28. Expected performance of the ATLAS b-tagging algorithms in Run-2. Technical report ATL-PHYS- PUB-2015-022. Geneva: CERN (2015). http://cds.cern.ch/record/2037697
29. Commissioning of the ATLAS high-performance b-tagging algorithms in the 7 TeV collision data. Technical report ATLAS-CONF-2011-102. Geneva: CERN (2011). https://cds.cern.ch/record/1369219
30. Aaboud M et al (2018) Measurements of b-jet tagging efficiency with the ATLAS detector using tt events at $\sqrt{s} = 13TeV$. JHEP 08:089. https://doi.org/10.1007/JHEP08(2018)089. arXiv: 1805.01845 [hep-ex]

31. Zyla PA et al (2020 (Particle Data Group). Review of particle physics. PTEP 2020(8):083C01. https://doi.org/10.1093/ptep/ptaa104
32. Reconstruction, Energy Calibration, and Identification of Hadronically Decaying Tau Leptons in the ATLAS Experiment for Run-2 of the LHC. Technical report ATL-PHYS-PUB- 2015-045. Geneva: CERN (2015). https://cds.cern.ch/record/2064383
33. Aaboud M et al (2018) Performance of missing transverse momentum reconstruction with the ATLAS detector using proton-proton collisions at $\sqrt{s} = 13TeV$. Eur Phys J C78(11):903. https://doi.org/10.1140/epjc/s10052-018-6288-9. arXiv: 1802.08168 [hep-ex]

# Chapter 6
# Fast Shower Simulation in the Forward Calorimeter

## 6.1 The ATLAS Simulation Infrastructure

The physics programs of the experiments at the LHC heavily rely on detailed simulations that predict the response of the complex detectors and make the comparison between experimental results and theoretical predictions possible. The simulation process in ATLAS is controlled by Athena [1], a software framework which takes care of the particle interaction and propagation in the detector, as well as the physics object reconstruction, described in Chap. 5. Athena uses the GEANT4 [2] simulation toolkit to propagate particles. The simulation chain can be divided into fundamental steps: the event generation, the detector simulation, the digitisation of the energy deposited in the sensitive material of the detector and finally the reconstruction of the physics objects [3]. The flow of the ATLAS simulation chain is sketched in Fig. 6.1, which reports the main stages (in boxes) and the data objects produced (arrows), and is briefly outlined below.

**Generation** The *pp* collision at the interaction point and the prompt decay of the emerging unstable ($c\tau < 10$ mm) particles are simulated with a Monte Carlo (MC) generator. Any stable particle that is expected to propagate through the detector is stored. The most commonly used MC generators in ATLAS are MADGRAPH [4, 5] and POWHEG [6] for matrix element generation, PYTHIA [7], SHERPA [8] and HERWIG [9] for parton shower and hadronisation. They typically produce simulation output files in the standard HepMC format. The output file contains information called *truth*, related to the history of the interactions from the generator for particles in the event.

**Simulation** The stable particles are propagated through the ATLAS detector by GEANT4. The structure of the detector is realistically reproduced and built from a database containing information about the detector geometry (volume dimensions, rotations, positions), as well as element and material properties (density, thickness). The simulation is performed in steps. For each step, GEANT4 is responsible for transporting the particle from one point to another. The length of each



C. Tosciri, *Higgs Boson Decays into a Pair of Bottom Quarks*, Springer Theses,
https://doi.org/10.1007/978-3-030-87938-9_6

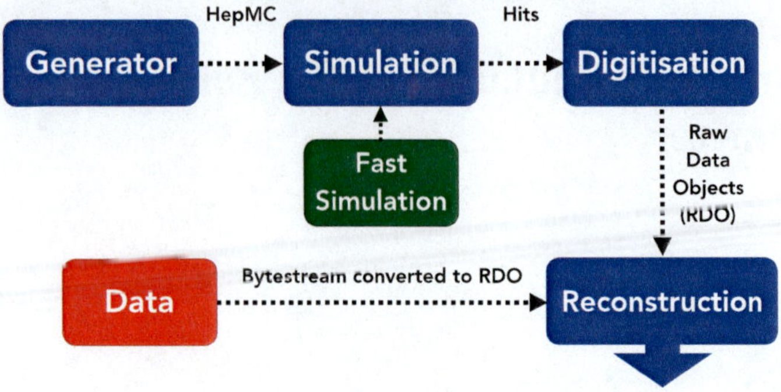


**Fig. 6.1** The flow of the ATLAS simulation software. Image adapted from [3].

step depends on the kinematic and the physical processes to which the particle is subject (like ionization), and on the macroscopic properties of the detector material traversed by the particle. The energy deposited in the sensitive portions of the detector is recorded as *hits*, containing the energy deposition, position and time. Truth-level information, like decays of specific particles or position of interaction, is also stored.

**Digitisation** The hits recorded in correspondence of the active material are converted into *digits*, which are used as input to the Read Out Driver (ROD), the readout module of the detector electronics. The ROD functionality is then emulated in order to produce a Raw Data Object (RDO) file. The ATLAS detector writes out data in a *bytestream* format and various algorithms are used to convert RDOs into bytestream and vice-versa. The truth information is also recorded at this level of the simulation chain.

**Reconstruction** The detector response is reconstructed into physical objects using various algorithms, described in Chap. 5. First, the response from the trackers and calorimeters are reconstructed as tracks and clusters, which are then combined into the physical objects such as jets, muons, electrons and photons. The results of the reconstruction are stored in a special format called Analysis Object Data (AOD). AODs can be read by the ATLAS analysis framework, which gives easy access to all the reconstructed objects. The truth information is processed up to the reconstruction step. This information can be used by the analyser to make studies at generator level. For instance, it could be possible to retrieve the origin of a specific simulated object, like a $b$-hadron that started a $b$-jet.

## 6.2 Fast Simulation

The complex and detailed simulation of physics events demands a large amount of computing resources which will dramatically increase with the increment of the luminosity at the LHC, expected with the Run-3 and the HL-LHC era. To address this limitation, considerable effort has been focused on the development of alternative and faster approaches for all the components of the full simulation chain [10]. Some examples of the tools used by ATLAS are the *FastCaloSim* [11], *FATRAS* [12] and *Fast Digitisation* [13] modules.

The most computationally demanding task in the chain is the detector simulation. In particular, for a typical physics event, most of the simulation time ($\sim$70%) is spent in the calorimeters, mainly in the end-cap and forward regions. The calorimeters are designed to fully absorb the particles in order to measure their energy and because of this, they are built of very dense material. From the simulation standpoint, the particle transport in a denser material requires a larger number of simulation steps, increasing the overall processing time. Since this is particularly true for the simulation of the FCal (described in Sect. 4.2.4), the present chapter is focused on developing novel methods to speed up the simulation of this sub-detector.

The FCal is a sampling calorimeter located at a distance of $\sim$5 m to the interaction point, covering the forward region in the range of pseudorapidity $3.1 < |\eta| < 4.9$. Because of its position, the FCal is exposed to a high flux of high-energy particles and their simulation demands a CPU consumption which increases exponentially with the multiplicity and the energy of the particles. Despite being not essential for most of the physics analyses in ATLAS, which usually select events in the low pseudorapidity region, an accurate simulation of the FCAL response is crucial for some important analyses. One example is the identification of the signal Higgs boson production in VBF processes [14]. To help the discrimination between QCD multi-jet events and VBF processes, the definition of certain event categories requires at least one VBF jet to be in the forward region, corresponding to the range $3.2 < |\eta| < 4.4$, which is fully covered by the FCal. The forward calorimeter is also fundamental for the measurement of the forward-backward asymmetry $A_{\text{FB}}$ which is sensitive to the electroweak mixing angle $\sin^2\theta_W$ and can be derived from cross section precision measurements [15].

The FCal is, therefore, a primary target to develop faster simulation methods by reducing the computational complexity while maintaining the needed accuracy for the response. Indeed, a fast simulation based on a *frozen shower library* approach for the FCal has been used by ATLAS since 2008 [16]. The approach is based on pre-simulated electromagnetic showers stored in a library and used later on to substitute low-energy particles in the simulation, at runtime. A specific frozen shower library must be generated for each type of particle and for each calorimeter that one wants to fast-simulate. In this work, the effort has been focused on the production and use of electromagnetic showers (in the following simply referred to as showers) generated in the first module of the forward calorimeter, FCal1, and initiated by electrons and photons, which are responsible for the majority of the shower production. Within

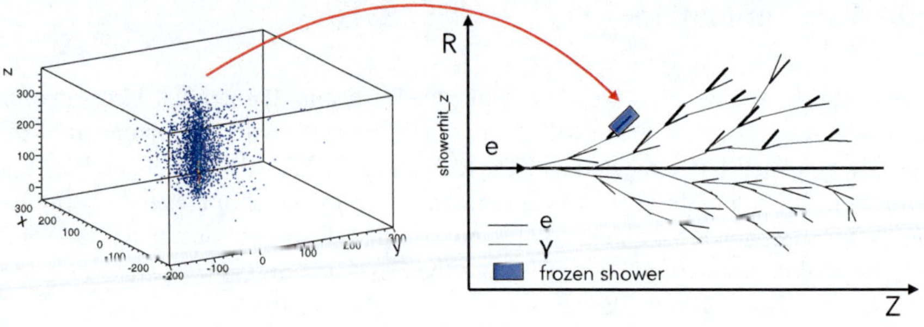


**Fig. 6.2** Sketch of the frozen shower substitution. On the left side, a shower (collection of hits) is shown as it is stored in the library. On the right, a low-particle produced in simulation is replaced with the stored shower

a sampling calorimeter, the energy of a showering particle is primarily deposited through a very large number of very soft electrons and photons. Several low-energy showers are, in fact, used for simulating a single high-energy particle entering the FCal, as sketched in Fig. 6.2.

A novel technique for the association of the showers stored in the library and the generated low-energy particles has been developed and will be described in the following sections. In Sect. 6.3, a general introduction to the fast simulation procedure based on the frozen shower approach is provided. In Sect. 6.4, the properties of the FCal are discussed as preliminary information for the description of the frozen shower libraries. A description of the default version of the library is given in Sect. 6.5, while Sect. 6.6 describes in details the new technique proposed and shows the relevant results.

## 6.3 Frozen Shower Library

The first step of the library generation is the simulation of so-called *starting points*, corresponding to the low-energy particles that initiate the electromagnetic showers to be stored in the frozen library. The simulation begins with top quark pair events $t\bar{t}$, generated at the interaction point. The stable particles produced in the decay of the $t\bar{t}$ pairs are propagated with GEANT4 until a particle with energy below a certain threshold is produced in the FCal, typically 1 GeV for electrons and 10 MeV for photons. The kinematic properties of these particles are saved as starting points in a library. The starting points are then propagated in the calorimeter until their whole energy is deposited, and the showers produced are stored as an aggregation of hits, in such a way that the library is a collection of starting points and their corresponding showers.

After the generation, the library can be used in the simulation. During the fast simulation, the detector response is simulated with GEANT4, until a particle of the same type of the library in use is produced below the energy cut-off. After some checks on the energy and position of the particle, to make sure that it is in the correct energy range and far enough from the edge of the calorimeter, the particle is matched with a given entry in the library. Before the substitution of the particle with the corresponding shower, since the energy of the matched shower will slightly differ from the original one, the energy of each hit in the shower is rescaled as indicated in Eq. (6.1):

$$E_{\text{hit}}^{\text{scaled}} = E_{\text{hit}} \times \frac{E_{\text{part}}}{E_{\text{part,lib}}}, \tag{6.1}$$

where $E_{\text{hit}}$ is the original energy of a single hit, $E_{\text{part}}$ is the energy of the eligible particle to be substituted and $E_{\text{part,lib}}$ is the energy of the corresponding particle in the library.

## 6.4 Properties of Electromagnetic Showers in FCal

The way a particle is matched with a shower in the frozen library depends on the characteristics of the development of the electromagnetic showers in the specific calorimeter. The structure of the FCal determines the parameters that should be used to identify the showers in the library.

As described in Sect. 4.2.4, the ATLAS forward calorimeter structure consists of an absorber matrix instrumented with cylindrical electrodes parallel to the beamline. The active material, the LAr filling the narrow interstices between each rod and tube, occupies a small portion of the calorimeter and is not distributed uniformly throughout its volume. In Fig. 6.3a, the rod structure of the FCal is reproduced by the transversal coordinates of the shower hits that provide the detectable signal. Only the energy released in the active material is, in fact, collectable in a sampling calorimeter. The *sampling fraction* indicates the ratio between the energy deposited by an incident particle in the active part of the detector and the total energy deposited, as described in Eq. (6.2) [17]:

$$f_{samp} = \frac{E(active)}{E(active) + E(absorber)}. \tag{6.2}$$

For the ATLAS FCal, the sampling fraction is about 1%. The calorimeter resolution depends on the sampling fraction, as well as on the *sampling frequency*, determined by the number of different sampling elements in the region where the showers develop. These two factors affect the *sampling fluctuations*, representing the variations of the shower energy deposited in the active calorimeter layers, from event to event. Sampling fluctuations are stochastic and they contribute to the total energy resolution, introduced in Eq. (4.6), as described by

$$\frac{\sigma_E}{E} \ni \frac{a}{\sqrt{E}}, \quad \text{with } a \propto \sqrt{\frac{d_{samp}}{f_{samp}}}, \tag{6.3}$$

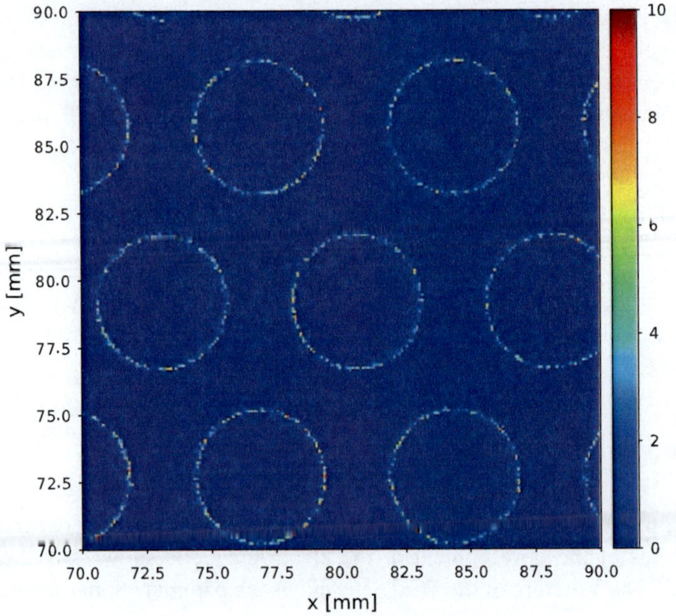


(a)

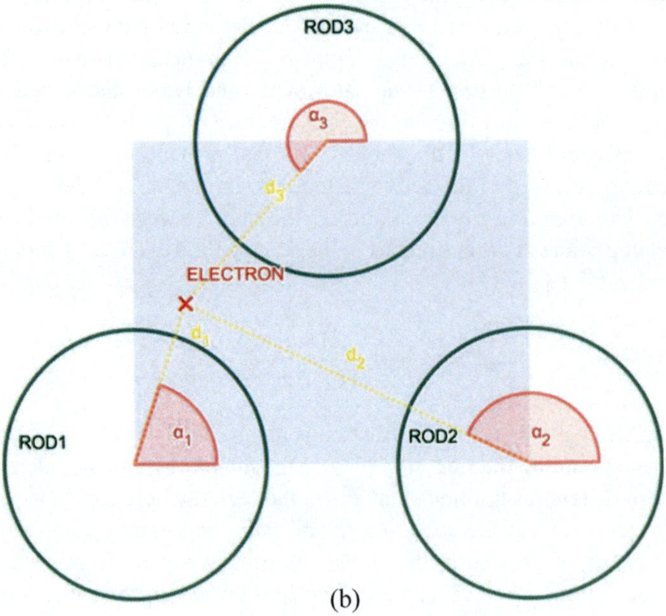


(b)

**Fig. 6.3** Transversal position of the shower hits collected in the FCal sub-detector (**a**). Scheme of three rods defining a unit cell in the structure of FCal: within the unit cell, the distance to the closest rod centre and the corresponding angle with respect to the $x$-axis are identified (**b**) in which $d_{samp}$ represents the thickness of individual active sampling layers [18, 19]. A small thickness means a large sampling frequency for a given total amount of active material and this translates into an increased number of particles contributing to the signal. On the other hand, the smaller the sampling fraction is, the larger is the stochastic term, as in the case of the FCal, where it represents a larger contribution compared to the ECAL and the HCAL. The dominant term in Eq. (4.6) for the FCal is the constant term. It includes contributions that do not depend on the energy of the particle and reflects non-uniformity of the detector response, mainly due to the special FCal geometry.

Due to the discontinuities between the active and the passive materials, the characteristics of the showers created in the FCal are expected to have a strong dependency on the position within the sub-detector element where the shower is originated. This position is identified as the distance $d$ to the centre of the closest rod and is defined as illustrated in Fig. 6.3b. Given an incident particle, the three closest rod are identified and the distance $d_{123}$ to the corresponding centres are calculated. The distance $d$ is defined as $d = min(d_1, d_2, d_3)$ and the angle $\alpha$ with respect to the $x$-axis is calculated accordingly. The non-uniformity of the FCal must be taken into account in the development of the algorithms used for simulation.

## 6.5 Default Library

The frozen libraries currently used by ATLAS for the FCal fast simulation (called *default libraries* in the following), contain about $10^4$ starting point-shower pairs, and are produced separately for electrons, photons and neutrons. For each library, the pre-simulated showers are parametrised in bins of position $d$ and pseudorapidity $\eta$ of the starting points, for which the energy is stored as well, but remains unbinned. For each shower, the position of the hits, their energy and time are saved. In 2017, a machine-learning-based procedure was introduced for defining an optimal binning in $d$ [20]. Before being stored in the library, each shower is translated and rotated to have the vertex corresponding to the interaction point at zero time, and momentum pointing to the positive $z$-axis. At simulation time, the starting point closest to the simulated low energy particle is found by first picking the closest $d$ and $\eta$ bin and then finding the closest energy available within the bin. After retrieving the needed shower, this must be translated to the spatial coordinates of the particle and rotated in the direction of its momentum.

The most critical aspect of the frozen showers approach is to reproduce the energy dependence of the resolution provided by the full simulation. Figure 6.4 shows that, with the default library, there are some problems with modelling the energy resolution, which is found to be two times smaller than the full simulation. Therefore, the parameters of the default library need to be tuned before being used. The tuning procedure provides reasonable agreement with the full simulation but in some regions of the phase space disagreements remain; moreover it is not automated and requires constant manual intervention.

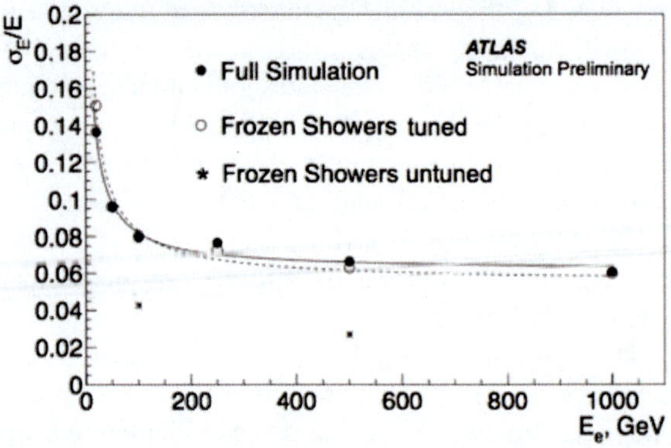


**Fig. 6.4** Energy resolution obtained in fast simulation using the default library (star points) compared to the result obtained with full simulation (black dots). The parameters in the library need to be tuned in order to obtain a result consistent with the full simulation. The empty circles correspond to the fast simulation response obtained with tuned library. *Image source* [20]

**Table 6.1** List of the features used in the default and new libraries to parametrise the starting points

| Feature | Default library | New library | Description |
|---|---|---|---|
| $\eta$ | Yes (discrete) | Yes | Particle rapidity |
| $d$ | Yes (discrete) | Yes | Distance to closest rod |
| $E$ | Yes | Yes | Particle energy |
| $\alpha$ | No | Yes | Angle to the closest rod |
| $\phi$ | No | Yes | Particle phi angle |
| $z$ | No | Yes | Particle $z$ coordinate |

## 6.6 Inverted Index Library

A new approach for the generation and use of frozen showers has been developed in order to overcome the limitations of the existing approach. The new version of the frozen library contains comprehensive information about the particle starting points: the stored showers are keyed by 6-dimentional parameter vectors containing the angular and spatial information of the starting points, as well as their energy. Table 6.1 summarises the features used to parametrise the starting points, in case of default and new library.

Figure 6.5 shows the distributions of the features describing the starting points of electrons generated with an energy threshold of 1 GeV, for the new library. The distribution of the distance to the closest rod, shown in Fig. 6.5a, has a triangular shape coming from the cylindrical symmetry of the rods. The distribution also shows a discontinuity in correspondence of the LAr gap. This is due to the different radiation lengths $X_0$ in argon (sensitive material, with $Z = 18$) and copper (absorber in FCal1, with $Z = 29$). The radiation length scales as

$$X_0 \propto \frac{A}{Z(Z+1)}, \tag{6.4}$$

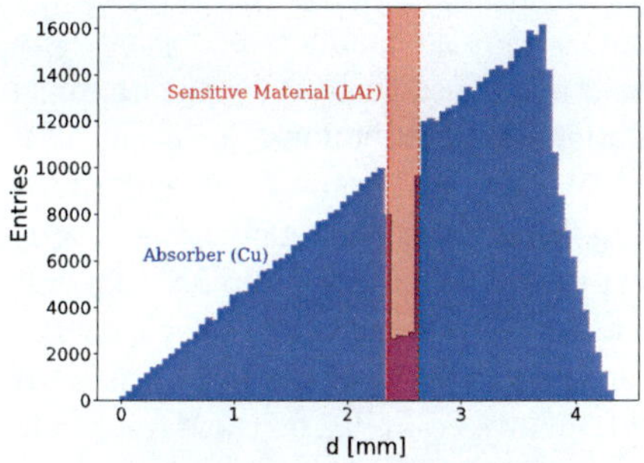


(a) Distance to the closest rod centre. The red band indicates the gap occupied by the sensitive material.

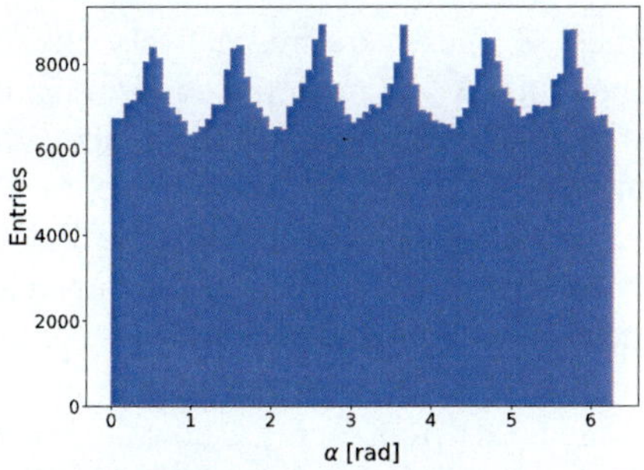


(b) Angle between the distance to the closest rod and the $x$-axis.

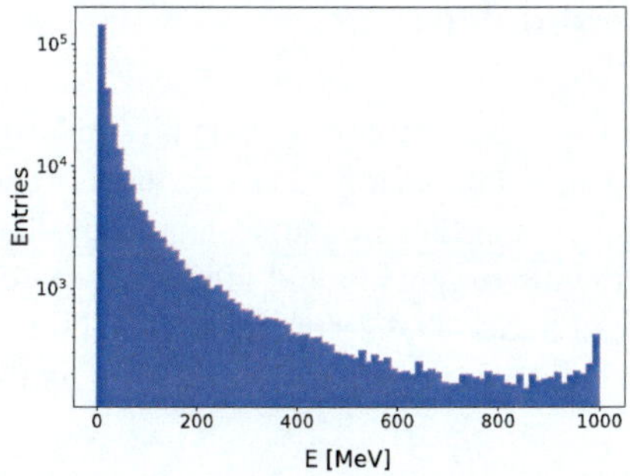


(c) Energy.

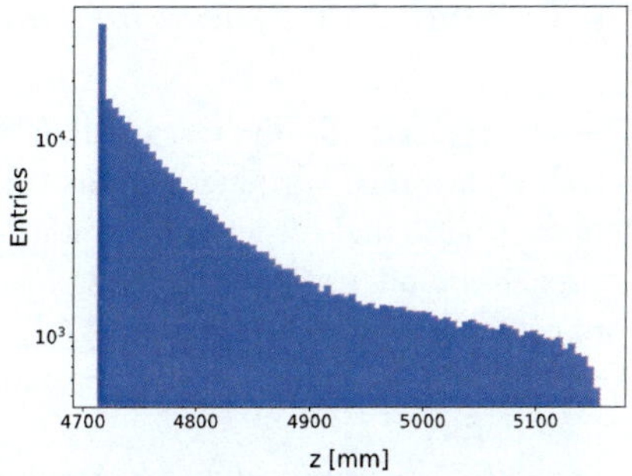


(d) Distance to the interaction point on the positive $z$-axis.

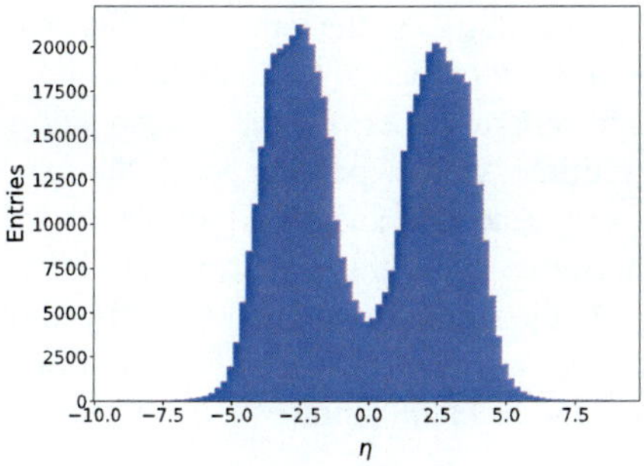


(e) Pseudorapidity.

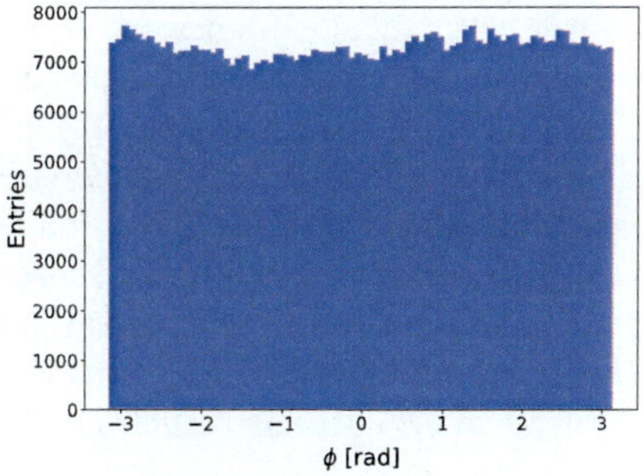


(f) Phi angle.

**Fig. 6.5** Distributions of the parameters describing electron starting points in the new library where $Z$ and $A$ are, respectively, the atomic and mass numbers of the nucleus. Therefore, $X_0$ is smaller in copper than in LAr and consequently, a smaller number of interactions takes place in the active material. The distribution of $\alpha$ in Fig. 6.5b, angle between the distance to the closest rod and the $x$-axis, presents regular spikes due to the use of polar coordinates for defining the angle in the rectangular geometry of the unit cell, identified by the three closest rods, as shown in Fig. 6.3b. Although the electrons initiating the showers are generated with a threshold of 1 GeV energy, they are very soft, as shown in Fig. 6.5c, and are most likely produced in the proximity of the FCal1 front face, as shown in Fig. 6.5d. Finally, from Fig. 6.5e and f, it can be argued that the starting points have uniform angular distributions.

### 6.6.1 Similarity Search for Fast Simulation

The new approach for the fast simulation of the FCal response is based on similarity search algorithms, introduced in Sect. 3.4. These are machine learning models that support search and retrieval of similar items in an efficient and scalable way. The approach includes several methods that allow us to quickly and efficiently search for vectors similar to each other. As described in Sect. 3.4.2, this requires to build a data structure, the index, from the database, which in this case is represented by the frozen library.

To this purpose, *Facebook AI Similarity Search* (Faiss) [21], a library developed by the Facebook Artificial Intelligence researchers to efficiently retrieve multimedia documents, has been used. Faiss is written in C++ and has been fully integrated in the ATLAS simulation framework. In a pre-processing step, the Faiss library is used to build the index which encapsulates the set of vectors for a given library. Then, the index structure is provided to the Athena framework, together with the corresponding library, and used during the simulation. At runtime, when a particle below the fixed energy threshold is produced, the simulation is suspended and such particle is used as a query vector to search for the closest starting point within the provided library. The similarity search is performed according to the similarity method chosen within Faiss. Once the closest starting point is found, the corresponding shower is returned to substitute the querying particle and finalise the fast simulation.

### 6.6.2 Indexing Methods in Faiss

Faiss allows us to focus the search to the most relevant part of the library, hence avoiding to examine a large portion of it. The vectors in the libraries are identified by an integer: the index. Since the similarity search performs better if all the objects have the same scale, the vectors have been normalised to the range [0, 1]. This operation is known in ML as 'feature scaling', which ensures that all the parameters are treated equally in the search process. To compare different vectors, a metric has to be defined, and while several metrics are available in Faiss, in the present work the Euclidean distance has been used. Different kinds of index have been described in Sect. 3.4.2. Faiss provides several methods which combine different indices and quantisers. The indexing algorithms learn the structure in a training phase. Vector compression is also available. Several combinations have been tested to find the most suitable one for showers retrieval. The Faiss methods explored in this work are briefly described below.

`IVF Flat` At training stage, Inverted File Index (IVF) clusters all input vectors into *nlist* groups using a $K$-means technique, as described in Sect. 3.4.2.1. The vectors are stored without compression. At search time, a number equal to *nprobe* of most similar groups to the query vector are returned and scanned exhaustively to provide the closest vectors. Only a fraction of the dataset is compared to the query: as a first approximation, it corresponds to the ratio *nprobe/nlist*, but this approximation is usually under-estimated because the inverted lists do not have equal lengths. A Flat index, which encodes the vectors into codes of a fixed size and store them in an array, is used as quantiser.

`HNSW Flat` The HNSW indexing method is based on a graph built on the indexed vectors, as described in Sect. 3.4.2.3. At search time, the graph is 'traversed' to find the nearest neighbours as quickly as possible. HNSW depends on a few parameters: $M$, the number of neighbours in the graph, *efConstruction* is the depth of exploration when the graph is being constructed, and *efSearch* is the depth of exploration of the search. A Flat index is used as quantiser.

`HNSW SQ` HNSW is used as index structure. The vectors are compressed to 8-bit integers using a Scalar Quantisation (SQ), which may cause some loss of precision.

`IVF HNSW` IVF index is used in the same way as IVF Flat, but the search function of the quantiser index is based on the HNSW graph structure.

`IVFPQ Flat` IVF index is used. Vectors are split into sub-vectors that are each quantised to 8 bits using a Product Quantisation (PQ) (see Sect. 3.4.2.2). A Flat index is used as quantiser.

`IVFPQ HNSW` IVF index is used with Product Quantisation (PQ). A quantiser based on HNSW graph structure is used.

### *6.6.3 Validation and Results*

For the validation of the new approach, several libraries with $\sim 10^5$ showers have been generated with electrons of energy below 1 GeV. Then, high-energy electrons are generated at the interaction point with ParticleGun, a tool in Athena, which allows us to "shoot" primary particles from a specific point with a certain energy, momentum and direction. The generated electrons have discrete values [100, 200, 300, 400, 500] GeV and pseudorapidity in the range $3.35 < |\eta| < 4.60$. They are propagated using the fast simulation with the new implementation. The validation returns the coordi-

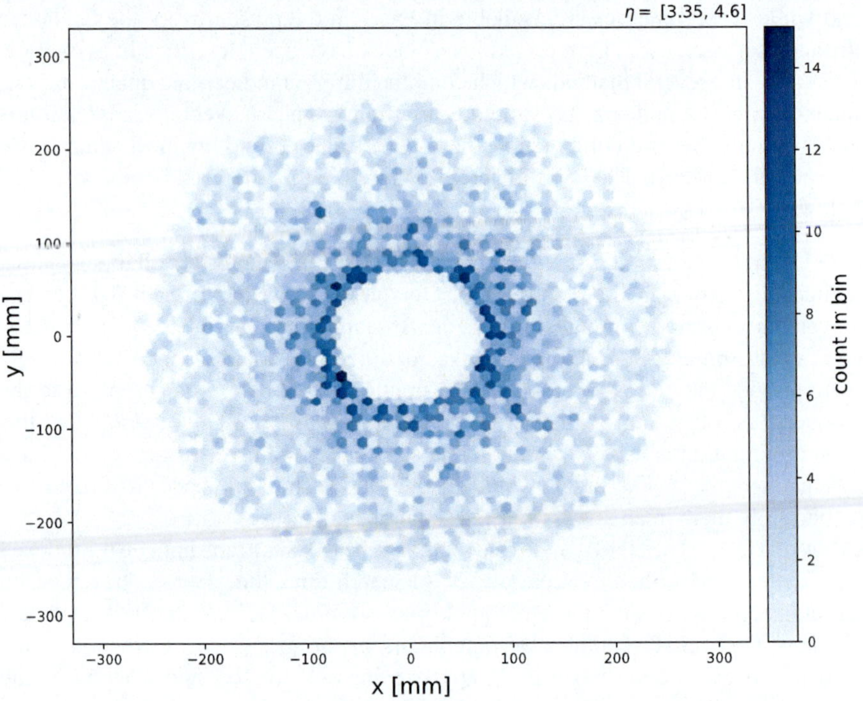


**Fig. 6.6** Spatial distribution of the generated high-energy electrons entering the FCal1 front face

nates of the FCal front face where the electrons interact, as well as the information related to the showers selected for the fast simulation. The validation is done both in the inclusive pseudorapidity range and for smaller pseudorapidity slices. Figure 6.6 shows the spatial distribution of the electrons entering the FCal1 front face.

The detector response is driven by the set of parameters shown in Table 6.1, which describe the starting points. However, this relation is stochastic and depends on the energy scale. The new method should provide good results in terms of CPU consumption and accuracy. These quantities can only be evaluated on the cumulative response of the simulation. The CPU consumption, expressed in seconds, is evaluated as the average of the CPU time, or processor time, spent executing the propagation of a high-energy electron of fixed initial energy. The propagation of a single high-energy electron produces a large amount of low-energy electrons, of the order of $\sim 10^4$, for each of which the search of the most suitable shower is performed. The accuracy of the method is evaluated from the detector resolution. This is measured as the ratio $\sigma_E/E$ between the standard deviation and the mean value of the energy deposited by the low-energy electrons in the calorimeter. The deposited energy corresponds to the sum of the energy of each hit of the low-energy showers, and its distribution is shown in Fig. 6.7 for the fully GEANT4-based simulation (blue), the default library (red) and the new library (green). As expected, the deposited energy depends on

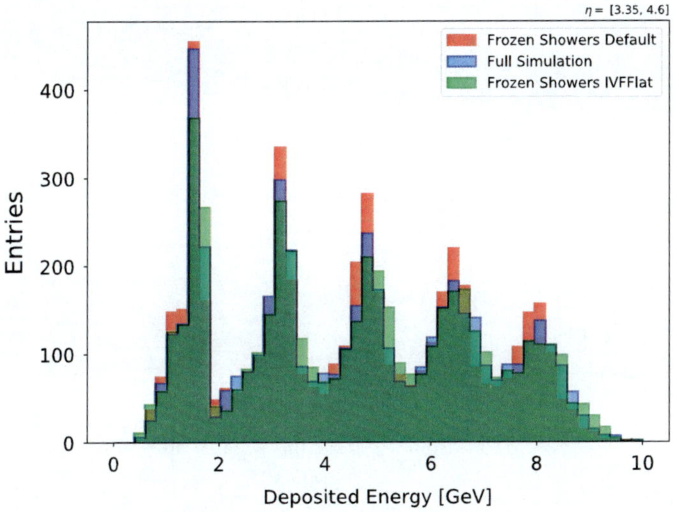


**Fig. 6.7** Distribution of the energy deposited by the showers originated by the low-energy electrons in the FCal calorimeter

the energy of the primary electron: the larger it is, the more secondary electrons and electromagnetic showers are generated, and the more energy is deposited in the calorimeter. Each peak in Fig. 6.7 corresponds to a specific discrete energy of the generated primary electrons, with an average deposited energy between 1 and 2% of the initial energy. This is due to the small sampling fraction of the FCal. Since the detector response is a stochastic process, the *bootstrap* method [22] has been used for assessing the statistical accuracy of the measure. The idea of the method is to randomly draw datasets with replacement from the original dataset for a fixed number of times (say 30 times). The original dataset here is represented by the sum of the energies deposited by the showers selected for each simulated event (i.e. each high-energy electron). For each bootstrap, the quantity of interest is computed, in this case the resolution $\sigma_E/E$, to obtain a distribution of such variable. From here, we can estimate the mean and the standard deviation of the variable of interest, as shown in Fig. 6.8.

#### 6.6.3.1 First Results (IVF Flat)

The validation of the new fast simulation procedure has been, at first, performed by using the `IVF Flat` index, the most common method used for similarity search and provided by Faiss. As shown in Fig. 6.9, the first results obtained with this method have been compared with the response provided by the full simulation (blue curves) of high energy electrons and with the response obtained with the default frozen library (red). The curves related to the `IVF Flat` search have been obtained by clustering

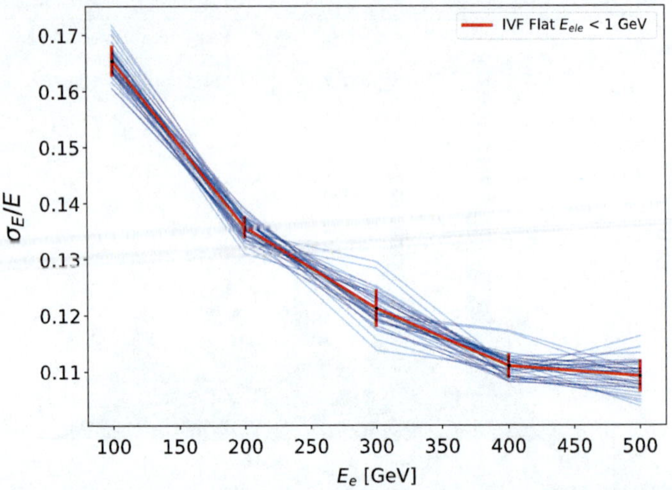


**Fig. 6.8** Detector resolution as a function of the energy of the initial electron. The estimation is obtained with the bootstrap method by sampling the initial dataset consisting of the output of the fast simulation

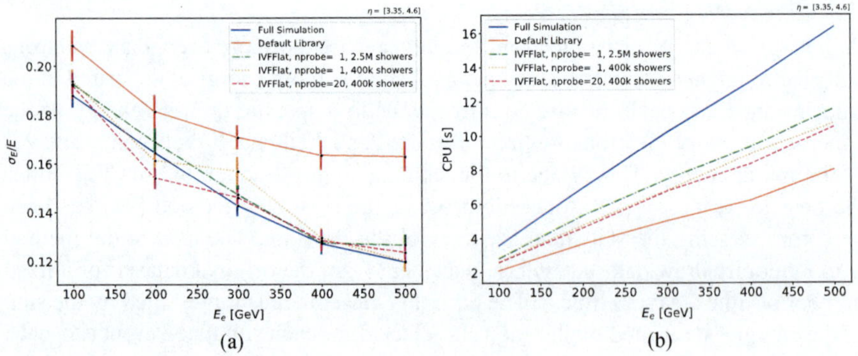


**Fig. 6.9** Resolution (**a**) and CPU consumption (**b**) resulting from the validation of the new fast simulation procedure, using the `IVF Flat` method. Results obtained with different libraries and different *nprobe* (green, yellow, purple) have been compared with Full Simulation (blue) and Default Library (red) results

the feature space with $nlist = 3000$, but using slightly different configurations. For the yellow and purple curves, the same library was used, containing $4 \times 10^4$ showers, but searching the closest shower within a different number of Voronoi regions, indicated by $nprobe$, equal to 1 and 20, respectively. The green curve was obtained with a larger library of $2.5 \times 10^5$ entries, whose clusters contain in average more showers. Figure 6.9a shows that the resolution obtained with the full simulation is always well reproduced by using the fast simulation with `IVF Flat` method. The new library outperforms the default one (red), which appears to have difficulties in

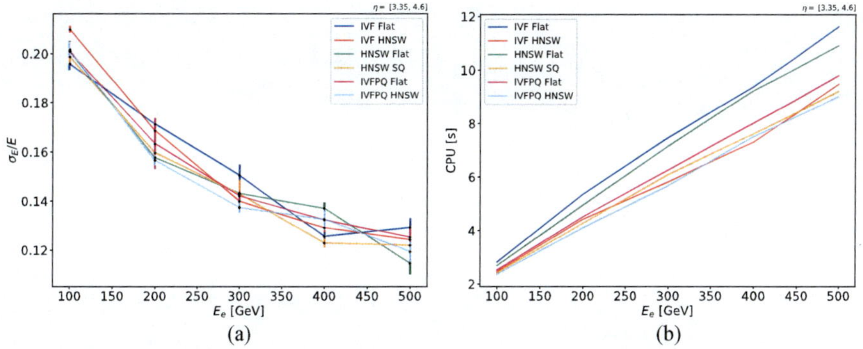


**Fig. 6.10** Resolution (**a**) and CPU consumption (**b**) resulting from the validation of the new fast simulation procedure, comparing different Faiss methods

reproducing properly the resolution of the full simulation. Figure 6.9b shows that the new approach also provides a faster simulation if compared to the full one. However, with the configurations shown in Fig. 6.9, `IVF Flat` is less efficient than the default fast simulation. There is not much difference between the three configurations: increasing *nprobe* from 1 to 20 does not introduce significant delay and does not provide better resolution. Increasing the library size, seems to provide a slightly better accuracy with, however, a small cost in the processing time.

#### 6.6.3.2 Comparison of Faiss Methods

The similarity search methods described in Sect. 6.6.2 have been tested in order to choose the most efficient one, making sure to keep the optimal resolution obtained with the `IVF Flat`. In Fig. 6.10, the results obtained with a validation in the inclusive pseudorapidity region are shown. From these results, different similarity search methods show quite analogous performances, with small improvements brought by the use of `IVF HNSW`, `IVFPQ HNSW` and `HNSW SQ`. However, looking at the performance in restricted regions of pseudorapidity, it can be realised that, in many cases, the IVF index built with an HNSW quantiser provides a faster search of the showers. This is clear from Fig. 6.11, which reports the CPU consumption in the ranges of $3.80 < |\eta| < 3.95$ (a) and $4.10 < |\eta| < 4.30$ (b). Given that the resolution is properly reproduced as well, the `IVF Flat HNSW` implementation has been selected as the optimal method to use for further studies of the fast simulation work.

#### 6.6.3.3 Optimisation

Both in the default library and the new one, each shower is stored with the vertex starting at the origin of the coordinate system, and oriented along the $z$-axis, as shown

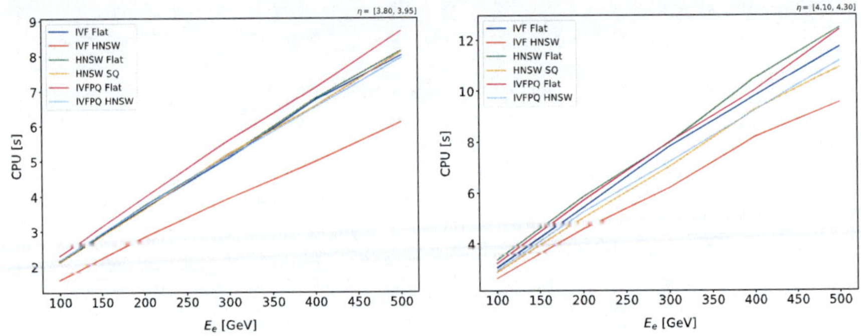


(a) CPU time in $3.80 < |\eta| < 3.95$. (b) CPU time in $4.10 < |\eta| < 4.30$.

**Fig. 6.11** CPU consumption resulting from the validation of the new fast simulation procedure in two pseudorapidity regions, comparing different Faiss methods

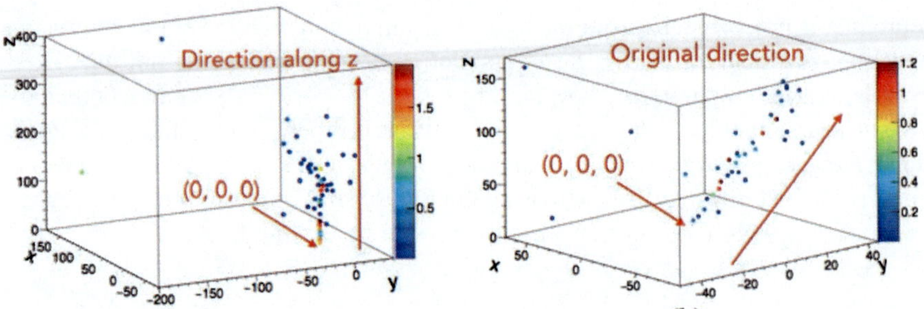


(a) (b)

**Fig. 6.12** Electromagnetic showers originated by electron starting points with energy $E < 1$ GeV. The shower in **a** is stored in the library after translation and rotation, while the shower in **b** is stored without receiving rotation

in Fig. 6.12a. At simulation time, when the closest shower has been selected, an affine transformation is performed to move the origin of the shower to the coordinates of the starting point and rotate it along its momentum. This is done because the default library does not use the $\phi$ angle for the shower search, and so the stored shower lying on the $z$-axis can then be rotated to appropriate $\phi$ direction (and $\eta$), once queried. This procedure is not required with the new approach and it can be dropped in order to speed up the simulation. The fact that the new library exploits the full set of particle parameters assures that the selected shower will have a direction which is likely to be very close to the direction of the particle momentum, making the two rotations unnecessary. Of course, the translation to make the shower corresponding to the starting point is still performed. Figure 6.12b shows that the shower rotation has been removed in such a way that the new library contains showers with origin in (0, 0, 0) and the original direction.

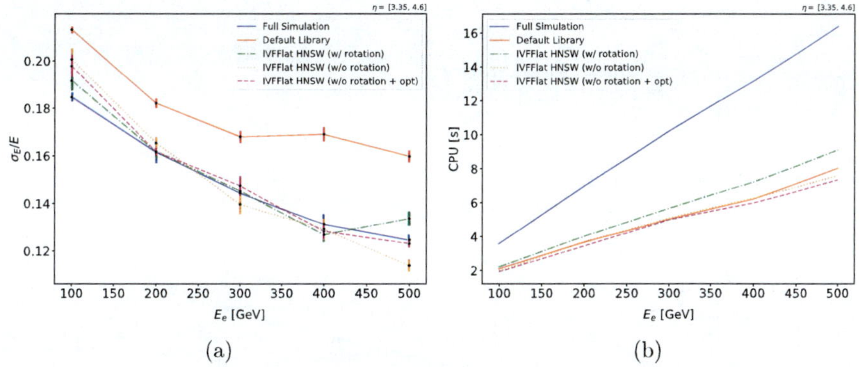


**Fig. 6.13** Resolution (**a**) and CPU consumption (**b**) resulting from the validation of the new fast simulation procedure, using the `IVFFlat HNSW` method

In Fig. 6.13, the green curves correspond to the results obtained with the `IVF Flat HNSW` algorithm with rotation of the showers, while the yellow one shows the results without rotation, as well as the purple curve, which includes, in addition, further optimisation of the relevant parameter *nlist* = *1000*. Figure 6.13b shows that removing the shower rotation provides a useful acceleration of the new simulation, which now outperforms the default electron library also in the computation consumption. The resolution obtained without rotation still reproduces nicely the full simulation response, as seen in Fig. 6.13a.

#### 6.6.3.4 Final Results

As mentioned, a dedicated library have to be generated for each particle type that needs to be fast-simulated. In addition to the library for the showers generated by electrons with energy $E_e < 1$ GeV, a new library has been produced with low-energy photon starting points. The energy threshold for the photons is 10 MeV. The choice of a different energy cut-off is due to previous studies which showed that very low-energy (below 10 MeV) photons do not create $e^+e^-$ pairs, making them suitable for the simulation. The photon library has been tested individually in the simulation and compared with the performance obtained with the default photon library. The comparison is shown in purple in Fig. 6.14: the full line represents the default library, while the dashed line represents the new photon library. The latter appears to be less efficient than the former and, as a general fact, the photon library provides a smaller boost to the simulation than the electron library. This is quite intuitive, given the lower threshold used for the photons. However, when the electron and photon libraries are used in combination, the simulation is significantly accelerated and provides a result comparable to the default library, as shown by the yellow curves. The difference in the individual photon performances is not reflected in the overall performance. This can be explained, again, by the fact that the thresholds used in the two libraries are

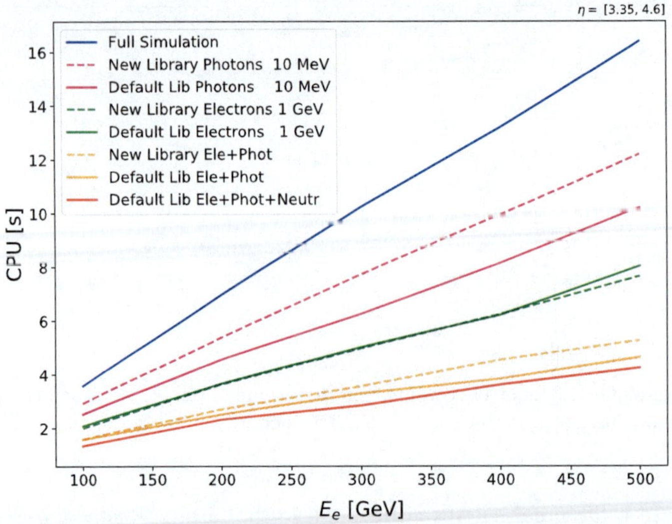


**Fig. 6.14** CPU response obtained with the new (dashed lines) and the old (full lines) libraries. The results come from simulations where the photon library (purple) and the electron library (green) are used individually for the fast simulation. In yellow is the response obtained with the simultaneous use of the two libraries. The results are compared with full simulation (blue) and the total default simulation which uses also a neutron library during the fast simulation (red)

very different: since the electrons have higher threshold, it is more likely that the corresponding library is used; additionally, every time that the electron library is called, the chance to use the photon library for lower energy photons is intrinsically killed.

For the sake of completeness, Fig. 6.15 shows the comparison of the performances of the two versions of the libraries. The new library approach (dashed curves) always provides a perfect reproduction of the full simulation response, while the default fast simulation (continuous curves) are less accurate.

## 6.7 Conclusions and Prospects

The new approach for the FCal fast simulation provides optimal results in terms of detector resolution response and improves the CPU time by around 70% with respect to the full simulation. The achieved acceleration is comparable to the default library method. This aspect can be further improved by performing the search of the most suitable showers in batches, since Faiss is optimised for batch searches. The main reason is that most index structures rely on a partition of the data that at query time requires a multiplication of a matrix with a vector, in case of single query, or another matrix, in case of multiple queries. Multiplications between matrices are

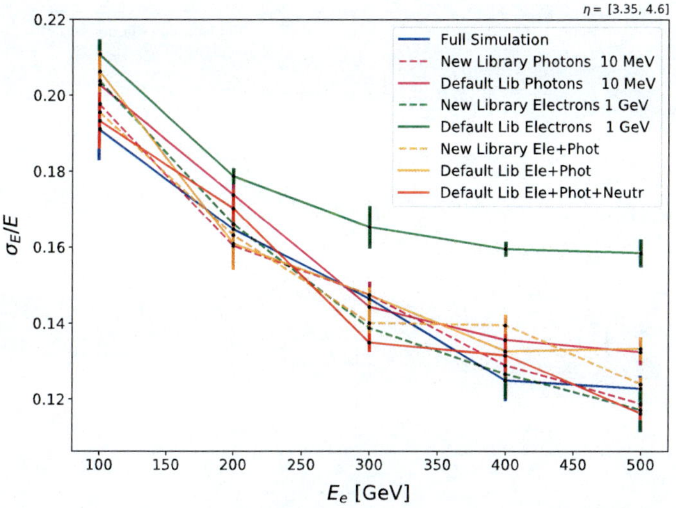


**Fig. 6.15** Resolution response obtained with the new (dashed lines) and the old (full lines) libraries. The photon library (purple), electron library (green) and their combination (yellow), both for the default and the new approach, are compared with the full simulation (blue) and the default simulation which uses a neutron library during the fast simulation (red)

usually much faster than the corresponding amount of matrix-vector multiplications. At the time when this work was finalised, the Athena simulation was implemented in such a way that only searches with single queries could be performed. The potential improvement due to batch searches has been tested using Faiss in a standalone mode. For this study two datasets (libraries) containing electron showers have been used. The first dataset, with $\sim 10^5$ showers, is used as the actual library. The second dataset contains $\sim 10^4$ showers, which are used as queries. A number of searches is performed on the first library using $2^n$ queries for each search, where $n = 0$ for the first group of searches, until all the queries of the dataset are matched. Then, $n$ is incremented by one and the searches are repeated to match exhaustively the entire dataset of queries. The procedure is repeated iteratively, incrementing $n$ by one at each step, until the size of the batch, $2^n$, corresponds to the size of the entire dataset and only one search is sufficient to match all the queries. For each value of $n$, the total CPU time required to match all the queries in the dataset is evaluated. Figure 6.16 shows that the CPU time decreases significantly as the batch size increases. This suggests that the fast simulation implemented in Athena would greatly benefit from batch searches.

In summary, the new technique outperforms the default approach and it can be used in production. Certainly, the similarity search library needs further validation. Comparisons between the results of the fast and full simulation for a variety of physics processes are necessary.

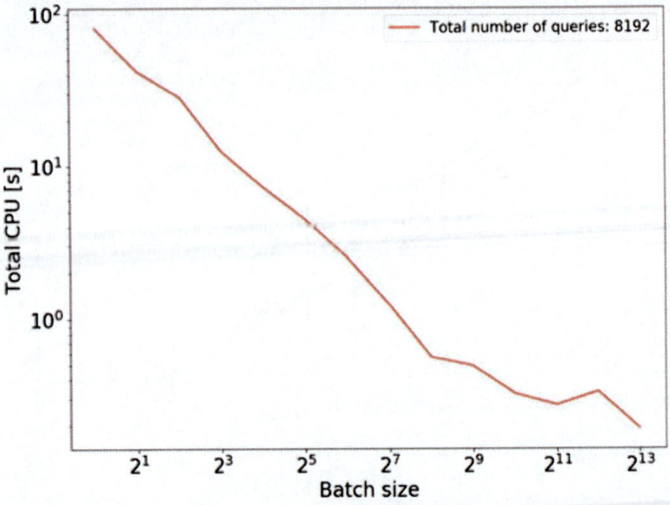


**Fig. 6.16** Total CPU time required by searches that exhaustively match a dataset of queries as a function of the batch size

## References


1. Calafiura P et al (2005) The Athena control framework in production. New developments and lessons learned. https://doi.org/10.5170/CERN-2005-002.456. https://cds.cern.ch/record/865624
2. Agostinelli S et al (2003) Geant4-a simulation toolkit. Nuclear Instrum Methods Phys Res Sect A: Acceler Spectr, Detect Assoc Equip 506(3):250–303. ISSN: 0168-9002. S0168-9002(03)01368-8. http://www.sciencedirect.com/science/article/pii/S0168900203013688
3. The ATLAS Collaboration et al (2010) The ATLAS simulation infrastructure. Eur Phys J C 70(3):823–874. ISSN: 1434-6052. https://doi.org/10.1140/epjc/s10052-010-1429-9
4. Alwall J et al (2011) MadGraph 5: going beyond. JHEP 06:128. https://doi.org/10.1007/JHEP06(2011)128. arXiv: 1106.0522 [hep-ph]
5. Alwall J et al (2014) The automated computation of tree-level and next-to-leading order differential cross sections, and their matching to parton shower simulations. JHEP 07:079. https://doi.org/10.1007/JHEP07(2014)079. arXiv: 1405.0301 [hep-ph]
6. Alioli S et al (2010) A general framework for implementing NLO calculations in shower Monte Carlo programs: the POWHEG BOX. JHEP 06:043. https://doi.org/10.1007/JHEP06(2010)043. arXiv: 1002.2581 [hep-ph]
7. Sjostrand T, Mrenna S, Skands PZ (2008) A brief introduction to PYTHIA 8.1. Comput Phys Commun 178:852–867. https://doi.org/10.1016/j.cpc.2008.01.036. arXiv: 0710.3820 [hep-ph]
8. Gleisberg T et al (2009) Event generation with SHERPA 1.1. JHEP 02:007. https://doi.org/10.1088/1126-6708/2009/02/007. arXiv:0811.4622 [hep-ph]
9. Bahr M et al (2008) Herwig++ physics and manual. Eur Phys J C58:639–707. https://doi.org/10.1140/epjc/s10052-008-0798-9. arXiv: 0803.0883 [hep-ph]
10. Basalaev A, Marshall Z (2015) The fast simulation chain for ATLAS. J Phys: Conf Ser 898:042016. https://doi.org/10.1088/1742-6596/898/4/042016. https://doi.org/10.1088%2F1742-6596%2F898%2F4%2F042016
11. Coll ATLAS et al (2010) The simulation principle and performance of the ATLAS fast calorimeter simulation FastCaloSim. Technical report. ATL-PHYS-PUB-2010-013. Geneva: CERN. https://cds.cern.ch/record/1300517

12. Mechnich J (2011) FATRAS - The ATLAS fast track simulation project. J Phys: Conf Ser 331:032046. https://doi.org/10.1088/1742-6596/331/3/032046
13. Jansky R (2015) The ATLAS fast Monte Carlo production chain project. J Phys: Conf Ser 664(7):072024. https://doi.org/10.1088/1742-6596/664/7/072024
14. Aaboud M et al (2018) Search for Higgs bosons produced via vector-boson fusion and decaying into bottom quark pairs in $\sqrt{s} = 13$ TeV pp collisions with the ATLAS detector. Phys Rev D 98:052003. https://doi.org/10.1103/PhysRevD.98.052003
15. The ATLAS collaboration et al (2015) Measurement of the forward-backward asymmetry of electron and muon pair-production in pp collisions at $\sqrt{s} = 7$ TeV with the ATLAS detector. J High Energy Phys 2015(9):49. ISSN: 1029-8479. https://doi.org/10.1007/JHEP09(2015)049
16. Barberio E et al (2015) Fast shower simulation in the ATLAS calorimeter. J Phys: Conf Ser 119(3):032008. https://doi.org/10.1088/1742-6596/119/3/032008
17. Fabjan CW, Gianotti F (2003) Calorimetry for particle physics. Rev Mod Phys 75:1243–1286. https://doi.org/10.1103/RevModPhys.75.1243
18. Livan M, Vercesi V, Wigmans R (1995) Scintillating-fibre calorimetry. CERN yellow reports: monographs Geneva: CERN. https://doi.org/10.5170/CERN-1995-002. https://cds.cern.ch/record/281231
19. Livan M, Wigmans R (2017) Misconceptions about calorimetry. Instru- ments 1. https://doi.org/10.3390/instruments1010003
20. Gasnikova K (2017) Measurement of the total W- and Z-boson production cross sections in pp collisions at $\sqrt{s} = 2.76$ TeV with the ATLAS detector. PhD thesis. DESY, Hamburg, Germany. https://edoc.hu-berlin.de/handle/18452/19981
21. Johnson J, Douze M, Jégou H (2017) Billion-scale similarity search with GPUs. CoRR. arXiv 1702:08734
22. Hastie T, Tibshirani R, Friedman JH (2009) The elements of statistical learning: data mining, inference, and prediction. Springer series in statistics. Springer, Berlin. https://web.stanford.edu/~hastie/ElemStatLearn/

# Chapter 7
# $VH, H \rightarrow b\bar{b}$ Search

The discovery of the Higgs boson in 2012 by the ATLAS and CMS collaborations [1, 2] was a major achievement in the quest to understand the electroweak symmetry breaking mechanism that generates the masses of elementary particles. Since then, substantial progress has been made to measure the properties of the Higgs boson with increased precision.

Most of its production and decay rates were measured and found to be consistent with the SM predictions. The gluon fusion and the vector boson fusion production modes and the Higgs decays into bosons $ZZ$ [3], $WW$ [4] and $\gamma\gamma$ [5], and fermions $\tau\tau$ [6] were all observed by the end of LHC Run-1. During Run-2, thanks to the higher luminosity available, it has been possible to measure additional fundamental pieces of the SM Higgs sector: the coupling with an up-type quark, through the measurement of the associated production with top quarks $t\bar{t}H$ [7], the coupling with a down-type quark, through the measurement of the decay mode $H \rightarrow b\bar{b}$, and the associated production mode with a vector boson $VH$ [8].

In this chapter, an overview of the $VH(H \rightarrow b\bar{b})$ analysis [8] that led to the observation of the SM Higgs boson decaying into pairs of bottom quarks $H \rightarrow b\bar{b}$, as well as the observation of the $VH$ production mode with the ATLAS detector at the LHC, is presented. After an introduction on the physics processes studied in the analysis, given in Sect. 7.1, the data and simulation samples needed and the event selection and categorisation used in the analysis are summarised in Sect. 7.2 and 7.3, respectively. The corrections applied to the $b$-jets are described in Sect. 7.4 with particular attention dedicated to the extensive studies performed in the context of the *muon-in-jet* correction. Section 7.5 is dedicated to the MVA description and studies. Sections 7.6 and 7.7 address, respectively, the statistical analysis and systematic uncertainties treatment. The results of the $VH(H \rightarrow b\bar{b})$ analysis are discussed in Sect. 7.8 and the cross-check analyses in Sect. 7.9. Finally, the combinations with other Higgs analyses, that provide the sensitivity for the observations, are summarised in Sect. 7.10.



C. Tosciri, *Higgs Boson Decays into a Pair of Bottom Quarks*, Springer Theses,
https://doi.org/10.1007/978-3-030-87938-9_7

## 7.1 Overview

The SM predicts that the Higgs boson decays into a $b\bar{b}$ pair with the highest branching ratio of $\text{BR}(H \to b\bar{b}) = 0.581 \pm 0.007\%$ [9]. Despite the large BR, selecting the $H \to b\bar{b}$ process among the copious multi-jet background produced in QCD interactions at the LHC is very challenging. The cross section of the dominant Higgs production mode, gluon fusion $ggH$, is about seven order of magnitude smaller than the cross section of QCD production of $b$-jets, as shown in Fig. 7.1. This makes the search of $H \to b\bar{b}$ very complicated in the $ggH$ channel. A powerful strategy is to focus on the mode where the Higgs boson is instead produced in association with a vector boson $V$, either a $W$ or a $Z$. The leptonic decays of $W$ and $Z$ provide signatures

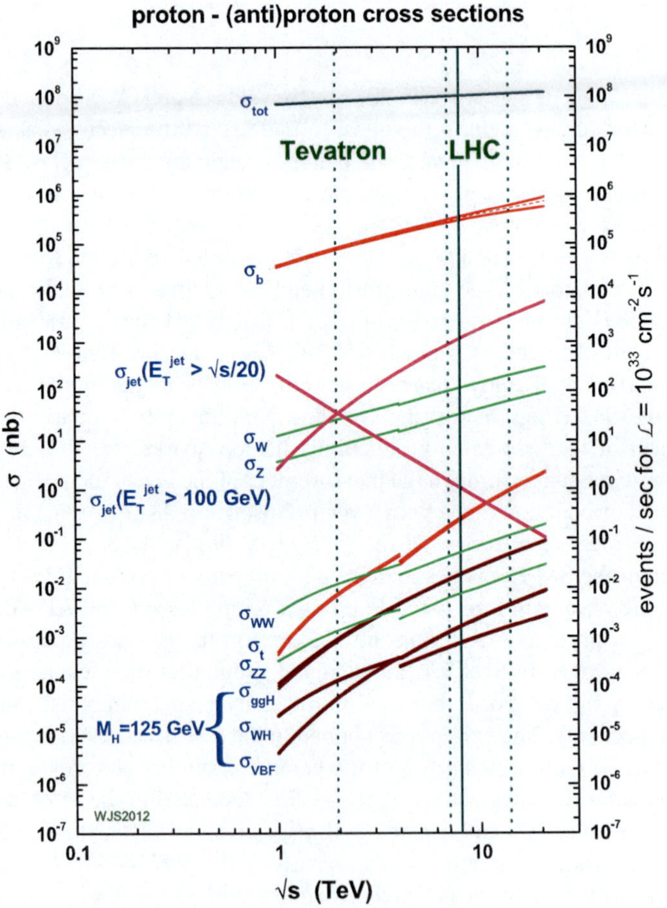


**Fig. 7.1** Standard Model cross sections of $pp$ and $p\bar{p}$ collision processes as a function of collider energy, with 125 GeV Higgs. *Image source* [10]

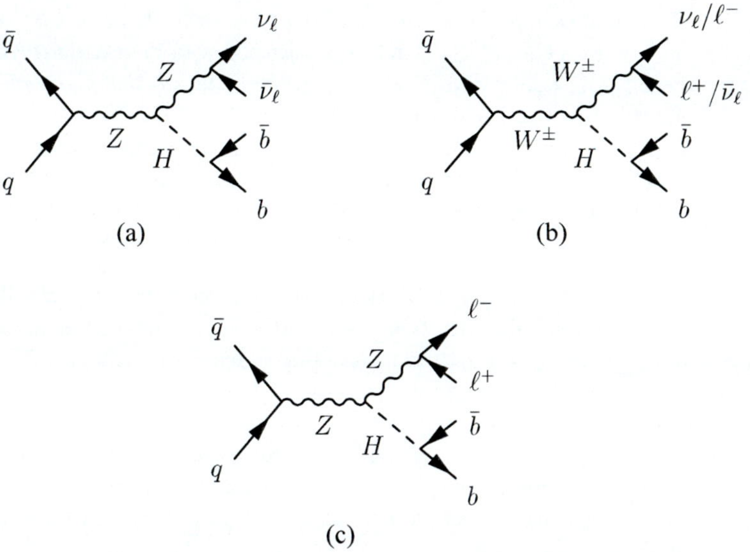


**Fig. 7.2** Feynman diagrams for the LO quark-initiated SM $VH(H \to b\bar{b})$ processes, with $V$ decaying leptonically

that allow for efficient triggering and powerful QCD background rejection. The $VH$ associated production is the most sensitive channel for the search of the $H \to b\bar{b}$ decay and was first exploited at the Tevatron with $p\bar{p}$ collision data. The combination of the CDF and D0 searches provided in 2012 the first evidence of the $VH(H \to b\bar{b})$ process with a global significance of 3.1 $\sigma$ [11]. A combination of Run-1 data at the LHC collected by the ATLAS and CMS experiments provided a significance of 2.6 $\sigma$ for the $H \to b\bar{b}$ decay in 2016 [6]. Finally, in Run-2 the ATLAS and CMS experiments separately announced the evidence of the Higgs boson decay into $b$-quark pairs and the associated production with a vector boson [12, 13]. Three different signatures are targeted in the corresponding analysis: $ZH \to \nu\nu b\bar{b}$, $WH \to \ell\nu b\bar{b}$ and $ZH \to \ell\ell b\bar{b}$, where $\ell = e, \mu$ and the vector boson decays into $\tau$-leptons are not explicitly considered. The corresponding Feynman diagrams are shown in Fig. 7.2.

## 7.2 Data and Simulation Samples

The $VH(H \to b\bar{b})$ analysis presented in this thesis was performed with the $pp$ collision data collected during Run-2 by the ATLAS detector at the LHC between 2015 and 2017, corresponding to an integrated luminosity of 79.8 $fb^{-1}$ at a centre-of-mass energy $\sqrt{s} = 13$ TeV. The majority of the SM backgrounds and the $VH(H \to b\bar{b})$ signal have been simulated with state-of-the-art MC simulators, such as POWHEG [14], PYTHIA [15] and SHERPA [16]. All samples of simulated events were generated at least at next-to-leading-order (NLO), passed through the ATLAS detector simulation based on GEANT4 [17] and reconstructed with the standard ATLAS software Athena [18], as described in Sect. 6.1. Both the quark- and gluon-induced processes contributing to the $ZH$ signal production have been taken into account. The mass of the Higgs boson was fixed at 125 GeV. The multi-jet background, produced by QCD processes, is largely suppressed by the selection criteria and the residual is evaluated using data-driven methods, as presented in Sect. 7.3.3. The remaining backgrounds involved in the measurement are briefly described in the following.

**V+jets production** A vector boson $V$ ($W$ or $Z$) is produced in association with jets. The main contribution is due to heavy-flavour (HF) jets. The jets containing heavy hadrons are tagged as $b$-jets, returning the same signature of the signal. $V + bb$, $V + bc$, $V + bl$ and $V + cc$ are jointly considered as $V$+HF background, which is dominant in several analysis regions.

**Top pairs production** Only leptonic and semi-leptonic decays of the $t\bar{t}$ pairs are included in this background. At least one lepton decaying from the $W$ bosons produced in the top decay is present in the final state, as well as heavy flavour jets mimicking the signal signature.

**Single top production** The total contribution of single top background, produced through $s$-channel, $t$-channel and $Wt$-channel, is smaller than $t\bar{t}$. This background has leptons and HF jets in the final state, arising from the top quark decay $t \rightarrow bW$.

**Diboson production** The diboson backgrounds are composed of three distinct processes $WZ$, $ZZ$ and $WW$, with the first two dominating. Both the $qq$- and $gg$-induced processes contribute to the final state.

Collectively, the background processes involving a vector boson decaying into leptons are referred to as electroweak (EW) backgrounds. Table 7.1 summarises the signal and background processes simulated for the analysis and also reports the corresponding theoretical predictions for the cross sections [19, 20], used to normalise the samples. The matrix element and parton shower generators used for each production are also indicated.

## 7.3 Selection and Categorisation

In order to target the three processes of interest $ZH \rightarrow \nu\nu b\bar{b}$, $WH \rightarrow \ell\nu b\bar{b}$ and $ZH \rightarrow \ell\ell b\bar{b}$, the events are categorised into three channels, 0-lepton, 1-lepton and 2-lepton, depending on the number of prompt electrons and muons in the final state. Since two $b$-jets are expected to be originated from the Higgs boson, a common requirement across the channels is the presence of exactly two $b$-tagged jets in the event. The presence of additional jets is allowed in some categories, which will be defined in the following of this section, according to the specific selection.

**Table 7.1** MC samples used for the signal and background processes with the corresponding cross section times branching ratio used to normalise the different processes at $\sqrt{s} = 13$ TeV. Branching ratios correspond to the decays shown. Table adapted from [19]

| Process | Generator | $\sigma \times$ BR [pb] |
|---|---|---|
| $qq \rightarrow Z(\nu\nu)H$ | POWHEG + PYTHIA | $8.91 \times 10^{-2}$ |
| $qq \rightarrow W(\ell\nu)H$ | POWHEG + PYTHIA | $26.91 \times 10^{-2}$ |
| $qq \rightarrow Z(\ell\ell)H$ | POWHEG + PYTHIA | $4.48 \times 10^{-2}$ |
| $gg \rightarrow Z(\nu\nu)H$ | POWHEG + PYTHIA | $1.43 \times 10^{-2}$ |
| $gg \rightarrow Z(\ell\ell)H$ | POWHEG + PYTHIA | $0.72 \times 10^{-2}$ |
| $Z \rightarrow \nu\nu+$ jets | SHERPA | 1914 |
| $W \rightarrow \ell\nu+$ jets | SHERPA | 20080 |
| $W \rightarrow \ell\ell+$ jets | SHERPA | $2107(m_{\ell\ell} > 40)$ GeV |
| $t\bar{t}$ | POWHEG + PYTHIA | 831.76 |
| single-top ($s$) | POWHEG + PYTHIA | 3.31 |
| single-top ($t$) | POWHEG + PYTHIA | 66.51 |
| single-top ($Wt$) | POWHEG + PYTHIA | 68.00 |
| $qq \rightarrow WW$ | SHERPA | 49.74 |
| $qq \rightarrow WZ$ | SHERPA | 21.65 |
| $qq \rightarrow ZZ$ | SHERPA | 15.56 |
| $gg \rightarrow VV$ | SHERPA | 56.27 |

### *7.3.1 Object Selection*

The physics objects are reconstructed consistently with the standard reconstruction procedures described in chapter 5. Lepton candidates must satisfy specific criteria of identification and isolation which define different working points.

By reference to Sects. 5.2.2 and 5.2.3, *loose electrons* are required to pass loose identification and track isolation requirements (*Loose ID* and *LooseTrack isolation*, respectively), while *tight electrons* must fulfil more stringent requirements (*Tight ID* and *HighPtCalo isolation*). In general, selected electron candidates have at least $p_{\mathrm{T}} = 7$ GeV, pseudorapidity within the ATLAS ID acceptance $|\eta| < 2.47$ and small impact parameter ($|z_0 sin\theta| < 0.5$ mm), to suppress pile-up contamination. Similarly, muon candidates are either selected with *loose* quality and isolation criteria (*loose muons*) or with *medium* quality and *tight* isolation (*tight muons*), according to the reconstruction requirements defined in Sects. 5.4.2 and 5.4.3. Muons are required to be within the acceptance of the muon spectrometer $|\eta| < 2.7$ and to have $p_{\mathrm{T}} > 7$ GeV and small impact parameter ($|z_0 sin\theta| < 0.5$ mm).

Jets are reconstructed with the anti-$k_t$ algorithm and radius parameter $R = 0.4$ (Sect. 5.5.1). In order to suppress pile-up, jets have to pass a requirement on the JVT discriminant (JVT $> 0.59$), introduced in Sect. 5.5.3. Jets in the central region $|\eta| < 2.5$ are required to have $p_{\mathrm{T}} > 20$ GeV, while for jets in the forward region ($2.5 < |\eta| < 4.5$) a stricter requirement of $p_{\mathrm{T}} > 30$ GeV is applied in order to further suppress jets from pile-up activity. The energies of the jets are calibrated using the standard calibration chain described in Sect. 5.5.2. $b$-tagged jets are subject to additional flavour-specific corrections aimed at improving their energy scale and resolution, as will be discussed in Sect. 7.4.

Since the complex reconstruction strategy has to deal with the signal left in several detector sub-systems, it is possible that the same signal is used for reconstructing different objects. To avoid any double-counting between leptons and jets, an *overlap removal* procedure is applied in the analysis, before the event selection. The main steps of the procedure are outlined in what follows:

- tau-electron: if $\Delta R(\tau, e) < 0.2$, the $\tau$ lepton is removed.
- tau-muon: if $\Delta R(\tau, \mu) < 0.2$, the $\tau$ lepton is removed.
- electron-muon: if a muon shares an ID track with an electron, the electron is removed if the muon is *combined*, instead the muon is removed if it is *calorimeter-tagged* (defined in Sect. 5.4.1).
- electron-jet: if $\Delta R(jet, e) < 0.2$, the jet is removed, otherwise if $\Delta R(jet, e) < \min\,(0.4, 0.04 + 10\ \text{GeV}/p_{\text{T}}^{e})$, the electron is removed.
- muon-jet: if $\Delta R(jet, \mu) < 0.2$, then the jet is removed if it has less than three associated ID tracks with $p_{\text{T}} > 500$ MeV or if the muon retains most of the jet energy, otherwise if $\Delta R(jet, \mu) < \min\,(0.4, 0.04 + 10\ \text{GeV}/p_{\text{T}}^{\mu})$, the muon is removed.
- tau-jet: if $\Delta R(\tau, jet) < 0.2$, the jet is removed.

The jets from hadronically decaying $\tau$ leptons are only used in the analysis in the overlap removal procedure.

**$b$-tagging**

In the analysis, jets are labelled as jets containing $b$-hadrons by imposing a cut-off on the MV2c10 discriminant, introduced in Sect. 5.5.4, consistent with a 70% efficiency measured in simulated $t\bar{t}$ events. The working point used in the analysis corresponds to a rejection rate (the inverse of the efficiency) of ~8 for $c$-jets and ~313 for light jets, as shown in Fig. 5.9b [21, 22]. A *direct tagging* is applied when the $b$-jet is tagged according to wether it passes the cut on the MV2c10 discriminant or not. However, while this is the unique approach for tagging jets in data, for some MC samples, removing the events that do not pass the MV2c10 selection would lead to a small selection efficiency that reflects on a poor statistics for the MC-based MVA training and fitting processes. Hence, a *truth tagging* can be applied in order to retain all the events of a sample and associate a weight to each MC event based on the expected probability for a jet to be tagged as a $b$-jet. The weight is calculated from the $b$-tagging efficiency of all the jets contained in the event. Truth tagging is used for the $V + cc$, $V + cl$, $V + ll$ and $WW$ samples, which simulate small background contributions. For all other samples, direct tagging is applied.

### 7.3.2 Event Selection

In all channels, events are required to have exactly two central $b$-tagged jets, arising from the decay of the Higgs boson candidate. The most energetic of the two $b$-jets is required to have $p_T > 45$ GeV. Events are categorised as *2-jet* if no additional untagged jets are found, or *3-jet* if such jets are present. In case of 0-lepton and 1-lepton channels, only one additional jet is permitted, since the events with four or more jets would add a large contamination from $t\bar{t}$ background against a small gain for the selected signal. On the contrary, in the 2-lepton channel, the $t\bar{t}$ contamination is limited by a special requirement on the invariant mass of the two leptons and the inclusion of high jet-multiplicity events provides a significant gain in the sensitivity (around 5% of improvement in the expected signal significance). Therefore, any number of jets is accepted in the 3-jet category of the 2-lepton channel.

Since the distribution of the vector boson transverse momentum $p_T^V$ for the signal is boosted at large values, as shown in Fig. 7.3 for 1-lepton and 2-lepton channels, the analysis focuses on the phase space with high $p_T^V$, with the aim of maximising the signal-to-background ratio. The $p_T^V$ considered, as well as specific selection requirements, optimised to reject the background, are described in the following.

**0-lepton channel** A missing transverse energy $E_T^{\text{miss}}$ trigger is used for the online selection, with threshold varying during the data-taking period from 70 GeV to 110 GeV, providing an efficiency between 85% and 90% at $E_T^{\text{miss}} = 150$ GeV [19]. Since this channel aims at selecting the $ZH \to \nu\nu b\bar{b}$ process, the vector boson transverse momentum $p_T^V$ is reconstructed as the $E_T^{\text{miss}}$ of the event and is required

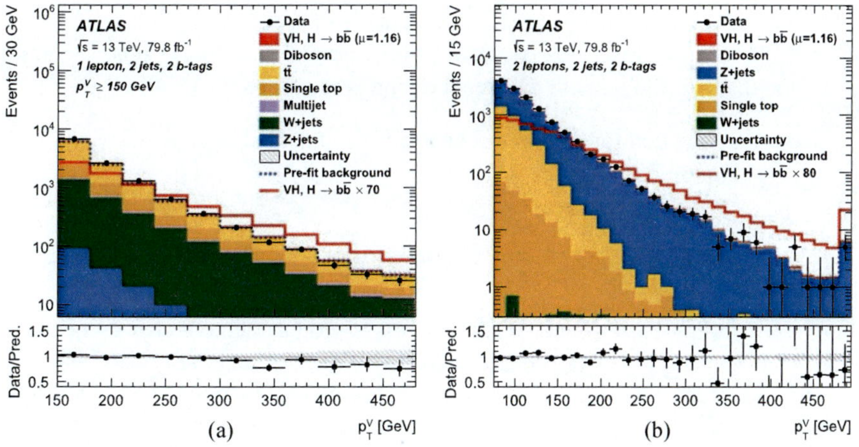


**Fig. 7.3** Distribution of the $p_T^V$ for the 2-jet category of 1-lepton (**a**) and 2-lepton channels (**b**). The distributions are obtained after performing the final fit to data. The Higgs boson signal and the backgrounds are normalised with the factors extracted in the fit and shown as a filled histograms. The signal is also shown unstacked as an unfilled histogram, scaled by the factor indicated in the legend. The dashed histogram shows the total pre-fit background. *Image source* [8]

to be larger than 150 GeV. A requirement on the scalar sum of the transverse momenta of the jets, $H_T$, is applied to remove a region which is mis-modelled in MC, due to a dependence of the trigger efficiency on the jet activity. No *loose* leptons are allowed in the events. In order to suppress the multi-jet background, a set of angular cuts are applied, as listed in Table 7.2 and explained in Sect. 7.3.3.

**1-lepton channel** Different online triggers are used for the two sub-channels. For the muon sub-channel, the same trigger as in 0-lepton channel is used, since in the analysis phase space it performs better than single-muon triggers. In the offline selection, a *tight* muon with $p_T > 25$ GeV and no additional *loose* leptons are required. In case of electron sub-channel, a single *loose* electron trigger is applied and, consequently, only events containing a single high $p_T$ *tight* electron are retained. The reconstructed vector boson transverse momentum $p_T^V$ corresponds to the vectorial sum of $E_T^{miss}$ and of the charged lepton transverse momentum. Across the channels, an additional selection of $E_T^{miss} > 30$ GeV is applied to reduce the background from multi-jet production, which is found to be non-negligible and is therefore modelled with a data-driven technique, as described in Sect. 7.3.3. The $p_T^V$ is required to be greater than 150 GeV. An additional selection separates a signal region, used to extract the signal in the statistical analysis, from a control region, used to constrain the background processes, as discussed in Sect. 7.3.4.

**2-lepton channel** The online selection is based on the presence of a single *loose* lepton. Then, an offline selection retains events with exactly two *loose* leptons with opposite charge, one of which must have $p_T > 27$ GeV and the invariant mass of the lepton pair compatible with that of the $Z$ boson. The multi-jet background is highly suppressed by this selection after which is found to be negligible. The $p_T^V$ is calculated as the transverse momentum of the 2-lepton system and required to be larger than 150 GeV for the high-$p_T^V$ region. Furthermore, the signal sensitivity is increased by the addition of a medium-$p_T^V$ category with 75 GeV $< p_T^V <$ 150 GeV. As described in Sect. 7.3.4, in this channel a control region that is orthogonal to the signal region is identified.

The event selection for the three channels is summarised in Table 7.2.

### 7.3.3 *Multi-jet Background Estimation*

In contrast to EW backgrounds, QCD processes produce leptons that do not arise from vector bosons directly produced in the collision, but rather originate from non-prompt weak decays, like semi-leptonic decays of heavy-flavour hadrons. These leptons may pass the trigger selection. Furthermore, mis-measurements of jets in the calorimeter can lead to non negligible $E_T^{miss}$ signature. The multi-jet background is treated with different methods across the channels, as discussed below.

**Table 7.2** Summary of the event selection and categorisation in the 0-, 1- and 2-lepton channels. *Table source* [8]

| Selection | 0-lepton | 1-lepton | | 2-lepton |
|---|---|---|---|---|
| | | $e$ sub-channel | $\mu$ sub-channel | |
| Trigger | $E_T^{miss}$ | Single lepton | $E_T^{miss}$ | Single lepton |
| Leptons | 0 *loose* leptons | 1 *tight* electron | 1 *tight* muon | 2 *loose* leptons with $p_T > 7$ GeV |
| | with $p_T > 7$ GeV | $p_T >$ 27 GeV | $p_T >$ 25 GeV | $\geq 1$ lepton with $p_T > 27$ GeV |
| $E_T^{miss}$ | $> 150$ GeV | $> 30$ GeV | – | – |
| $m_{\ell\ell}$ | – | – | | 81 GeV $< m_{\ell\ell} <$ 101 GeV |
| Jets | Exactly 2 / Exactly 3 jets | | | Exactly 2 / $\geq 3$ jets |
| Jet $p_T$ | $> 20$ GeV for $\lvert\eta\rvert < 2.5$ | | | |
| | $> 30$ GeV for $2.5 < \lvert\eta\rvert < 4.5$ | | | |
| $b$-jets | Exactly 2 $b$-jets | | | |
| Leading $b$-jet $p_T$ | $> 45$ GeV | | | |
| $H_T$ | $> 120$ GeV (2 jets), $>150$ GeV (3 jets) | – | | – |
| $\min[\Delta\phi(\mathbf{E}_T^{miss}, \vec{jets})]$ | $> 20°$ (2 jets), $> 30°$ (3 jets) | – | | – |
| $\Delta\phi(\mathbf{E}_T^{miss}, \vec{bb})$ | $> 120°$ | – | | – |
| $\Delta\phi(\vec{b_1}, \vec{b_2})$ | $< 140°$ | – | | – |
| $\Delta\phi(\mathbf{E}_T^{miss}, \vec{p}_T^{miss})$ | $< 90°$ | – | | – |
| $p_T^V$ regions | $> 150$ GeV | | | 75 GeV $< p_T^V <$ 150 GeV, $> 150$ GeV |
| Signal regions | – | $m_{bb} \geq 75$ GeV or $m_{top} \leq 225$ GeV | | Same-flavour leptons Opposite-sign charges ($\mu\mu$ sub-channel) |
| Control regions | – | $m_{bb} < 75$ GeV and $m_{top} > 225$ GeV | | Different-flavour leptons Opposite-sign charges |

**0-lepton**

The multi-jet background enters the 0-lepton selection because of jet mis-measurements. The fake $E_{\mathrm{T}}^{\mathrm{miss}}$ is expected to be aligned with the mis-measured jet, while a correctly reconstructed $E_{\mathrm{T}}^{\mathrm{miss}}$, representing the $Z$ boson, would roughly be directed in the opposite direction. To suppress the multi-jet background, a set of angular cuts is applied as follows:

- $\min[\Delta\phi(\mathbf{E}_{\mathrm{T}}^{\mathrm{miss}}, \vec{\mathrm{jets}})] > 20°(2\ \mathrm{jets}), > 30°(3\ \mathrm{jets})$
- $\Delta\phi(\mathbf{E}_{\mathrm{T}}^{\mathrm{miss}}, \vec{bb}) > 120°$
- $\Delta\phi(\vec{b_1}, \vec{b_2}) < 140°$
- $\Delta\phi(\mathbf{E}_{\mathrm{T}}^{\mathrm{miss}}, \vec{p}_{\mathrm{T}}^{\mathrm{miss}}) < 90°$

In the listed conditions, $\phi$ is the azimuthal angle, $\vec{b_1}$ and $\vec{b_2}$ are the two selected $b$-jets arising from the Higgs candidate, $\mathbf{E}_{\mathrm{T}}^{\mathrm{miss}}$ is the missing energy transverse of the event and $\vec{p}_{\mathrm{T}}^{\mathrm{miss}}$ is the missing transverse momentum calculated using only tracks reconstructed in the inner tracking detector and matched to the primary vertex. Fake missing energy $\mathbf{E}_{\mathrm{T}}^{\mathrm{miss}}$ tends to diverge from the track-based transverse momentum $\vec{p}_{\mathrm{T}}^{\mathrm{miss}}$. After the angular selection, the presence of multi-jet events is considered to be negligible.

**1-lepton**

The multi-jet background is not negligible in the 1-lepton channel and it can not be extracted using simulations, because of insufficient MC statistics due to the large cross section of QCD processes. Therefore, it is estimated with a template fit performed separately for the electron and muon sub-channels in 2-jet and 3-jet categories. A multi-jet enriched control region is constructed from data by requiring an inverted lepton isolation selection, in such a way that the control region contains events that are kinematically similar to the events of the corresponding signal region, but do not overlap with them. To increase the statistics of the background estimation, the requirement on the number of $b$-tagged jets in the final state is relaxed to one. After subtracting the contribution from EW background processes, evaluated with MC predictions, the multi-jet yield is extracted. Then, a fit to the $W$ boson candidate's transverse mass, reconstructed as $m_{\mathrm{T}}^{\mathrm{W}} = \sqrt{2p_T^{\ell} E_{\mathrm{T}}^{\mathrm{miss}}(1 - cos(\Delta\phi(\ell, E_{\mathrm{T}}^{\mathrm{miss}})))}$, is performed in the signal region to determine the normalisation of the multi-jet contribution, which is used in the final global likelihood fit. The $m_{\mathrm{T}}^{\mathrm{W}}$ variable is chosen because it offers an optimal discrimination between QCD- and EW-induced processes. The multi-jet contribution in the 2-jet category is found to be 1.9%(2.8%) of the total background contribution in the electron (muon) sub-channel, while in the 3-jet category is found to be 0.2%(0.4%).

**2-lepton**

The multi-jet background in the 2-lepton channel is efficiently suppressed by the dilepton invariant mass requirement: 81 GeV $< m_{\ell\ell} <$ 101 GeV. Any residual contribution is evaluated with a template fit to the dilepton mass distribution, where the expected EW contributions evaluated in MC are also included. The fraction of the remaining background coming from multi-jet events is estimated to be small enough to have a negligible impact on the signal extraction, therefore it is not included in the global likelihood fit.

### 7.3.4 Analysis Regions

Events passing the selection criteria are classified in eight categories, corresponding to the signal regions (SR) of the analysis. Additional categories, defined as control regions (CR) and relying on different selections, are constructed in order to constrain some background processes such as the $W$ boson production in association with jets containing heavy-flavour hadrons ($W$+HF CR) and top quark pair production ($e\mu$ CR). CRs do not overlap with the SRs and are characterised by negligible signal contamination. The two CRs are described in what follows.

**$W$+HF** It is defined from the 1-lepton channel both in the 2-jet and 3-jet high-$p_T^V$ categories to be highly pure in $W$+HF. It is built by applying, on top of the 1-lepton selection, a cut on the reconstructed top mass $m_t$, which helps to reduce the $t\bar{t}$ and single top background, and a cut on the invariant mass of the dijet system $m_{bb}$, to reduce the $VH$ signal contamination. The specific cuts are $m_{top} > 225$ GeV and $m_{bb} < 75$ GeV. The resulting purity of the $W$+HF control region is around 75%. In Fig. 7.4 the event yield in the $W$+HF CR is shown for 2-jet (a) and 3-jet (b) categories.

**$e\mu$** It is constructed by requiring the same kinematic selection as 2-lepton signal region, but different flavours for the two leptons ($e\mu$ or $\mu e$), which is not an expected signature for $ZH$. The assumption is that the kinematics of the $t\bar{t}$ same-flavour SR and the $e\mu$ CR are the same, since $t\bar{t}$ is flavour-symmetric. This was proved in the first Run-2 analysis [23] by comparing the shape of the observables in the two regions. This CR achieves a 99% purity in $t\bar{t}$ and single top. To constrain the top background yield and shape, the $m_{bb}$ distribution, which is one of the most sensitive discriminants, is used in the statistical analysis. The $m_{bb}$ distribution in the $e\mu$ CRs is shown in Fig. 7.4 for the high-$p_T^V$ region in 2-jet (7.4c) and 3-jet (7.4d) categories. The different binning of the two distributions is due to the limited statistics in the 2-jet high $p_T^V$ region. The $e\mu$ CRs are also evaluated and used in the medium-$p_T^V$ regions.

In the analysis, a total of eight SRs and six CRs are defined, as summarised in Table 7.3. After the selection, the three channels show a very diverse compositions of the backgrounds: the top quark pairs production ($t\bar{t}$) is dominant in 1-lepton and also affects 0-lepton and 2-lepton channels; $Z$+jets is the main background in 0-lepton and 2-lepton; $W$+jets contributes to both 1-lepton and 0-lepton, while does not affect the 2-lepton channel because of the $m_{\ell\ell}$ constrain; minor backgrounds such as single top and vector boson pairs production (Diboson) contribute in small proportions to all channels; the multi-jet background represents a small contribution, only in 1-lepton channel. In Fig. 7.5, the invariant mass distribution of the two $b$-

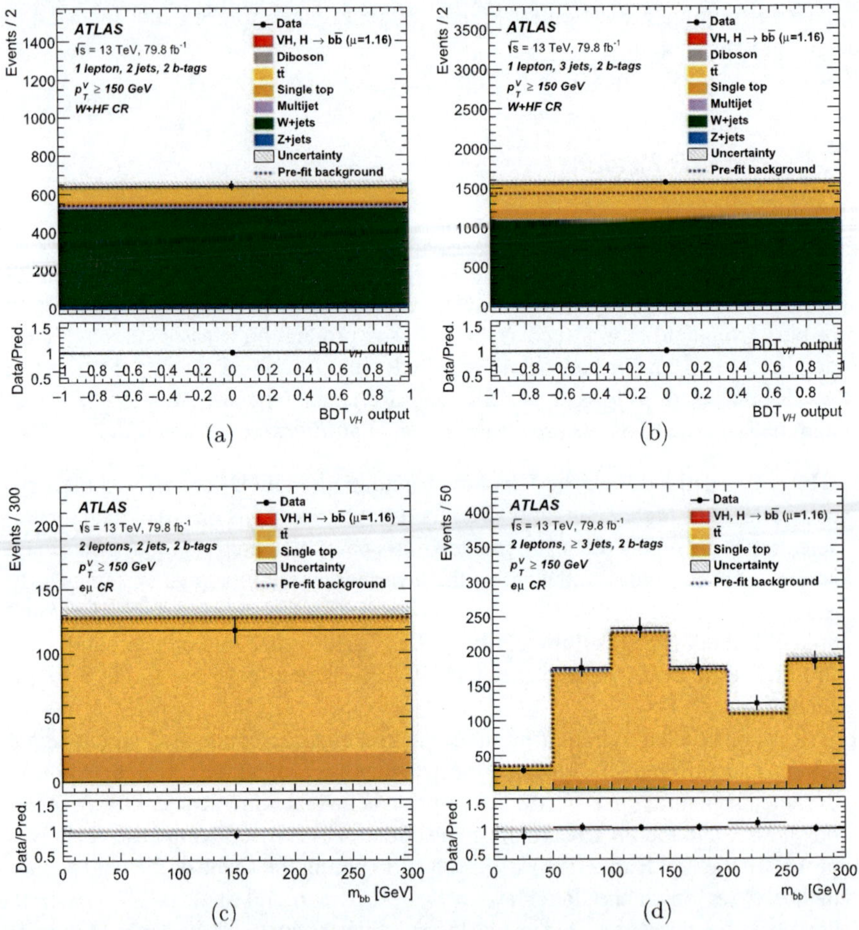


**Fig. 7.4** Control region distributions used as input to the global likelihood fit for the MVA analysis to extract the normalisation of the main backgrounds. *Image sources* [8]

jets is shown for signal and backgrounds to highlight the different compositions in 0-lepton (7.5a), 1-lepton (7.5b) and 2-lepton (7.5c).

## 7.4 *b*-jet Energy Corrections

As introduced in Sect. 5.5.2, the energy of the jets in ATLAS is calibrated using a series of standard corrections derived with MC and data-driven techniques. These corrections do not account for differences in jet flavours. However, heavy-flavour jets present some peculiarities that can be exploited not only for the identification

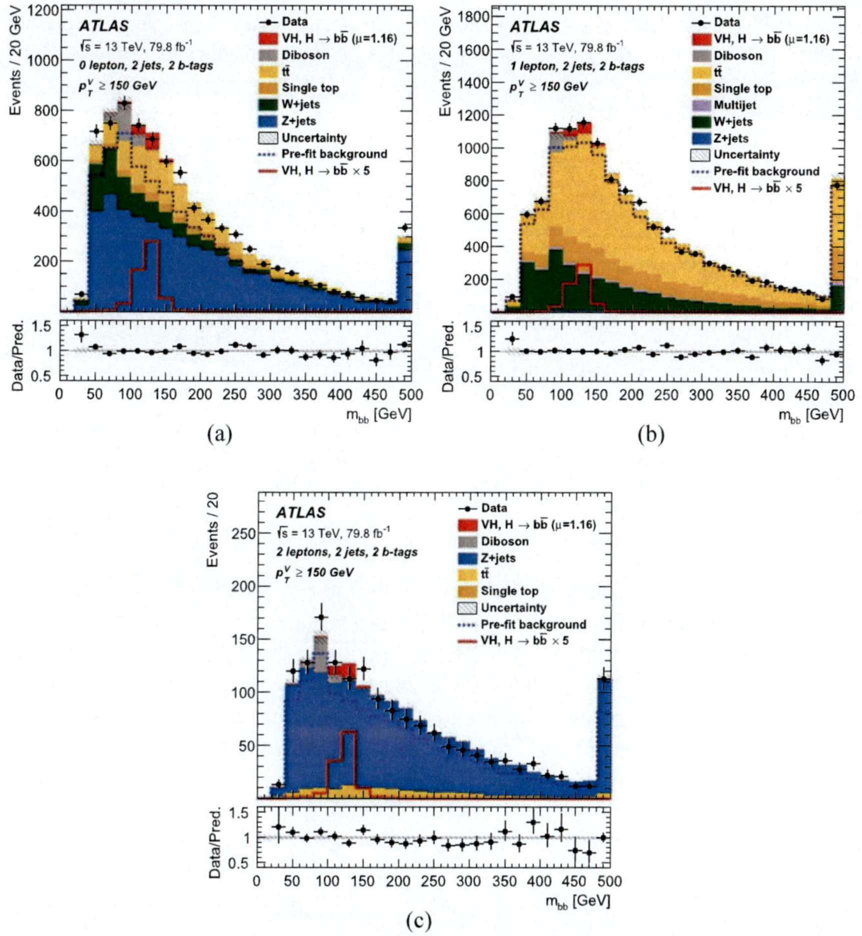


**Fig. 7.5** Invariant mass distribution of the two *b*-jets in 0-lepton (**a**), 1-lepton (**b**) and 2-lepton (**c**) channels. The distributions have been obtained after performing the final likelihood fit, therefore the extracted normalisation factors are applied. The dashed histogram shows the total pre-fit background. The background contributions are shown as filled histograms. *Image sources* [8]

of the jet flavour (as in the case of displaced vertices used as a precious information for *b*-tagging), but also to further correct the energy of the jet, on top of the standard ATLAS jet corrections.

Specific corrections for *b*-jets account for semileptonic decays of the *b*-quark occurring in the *b*-hadrons contained in the jet. The dominant decay mode of a *b*-quark is $b \rightarrow cW^*$, where the virtual *W* materialises either into a pair of leptons $\ell\nu$ (*semileptonic decay*) or into a pair of quarks which then hadronise (*hadronic decay*) [24]. An additional virtual *W* arises from the decay of the *c*-quark produced in the *b*-quark decay, with the possibility to produce an extra $\ell\nu$ couple in the *b*-

**Table 7.3** Summary of the analysis regions. Regions with relatively large signal purity are marked with the label SR. Regions designed to constrain the background uncertainties are marked with the label CR. *Table source* [8]

| Channel | Categories | | | | | |
|---|---|---|---|---|---|---|
| | 75 GeV < $p_T^V$ < 150 GeV | | | $p_T^V$ > 150 GeV | | |
| | 2jet | 3jet | ≥ 3jet | 2jet | 3jet | ≥ 3jet |
| 0-lepton | – | – | – | SR | SR | – |
| 1 lepton | | | | | | |
| $m_{bb} \geq 75$ GeV or $m_{top} \leq 225$ GeV | – | – | – | SR | SR | – |
| $m_{bb} < 75$ GeV and $m_{top} > 225$ GeV | – | – | – | CR | CR | – |
| 2-lepton | | | | | | |
| $ee$ and $\mu\mu$ channels | SR | – | SR | SR | – | SR |
| $e\mu$ channel | CR | – | CR | CR | – | CR |

quark decay chain. Examples of quark-level diagrams representing two possible $b$-jet semileptonic decays are shown in Fig. 7.6. For $b$-jets containing semileptonic decays of the $b$-quark, a large fraction of the jet momentum is carried by the leptons and neutrinos. While the energy of the electrons deposited in the calorimeter is included in the total energy of the jet, muons and neutrinos are not properly accounted for in the jet reconstruction. In fact, muons are detected in the MS and only a small fraction of their energy is deposited in the calorimeter, while neutrinos totally escape detection. This leads to a mis-calibration of the jets, yielding a jet energy response different from unity, with larger deviations at low jet $p_T$, called *non-closure* effect.

Since the signal of the $VH(H \to b\bar{b})$ analysis contains two $b$-jets originating from the Higgs resonance, a set of $b$-jet corrections is applied to improve the energy scale and resolution of these jets. The resolution of the reconstructed invariant mass $m_{bb}$, which provides a powerful discrimination between signal and background, widely benefits from the $b$-jet corrections, resulting in a significant gain in the sensitivity of the analysis. In particular, two consecutive corrections are applied to $b$-tagged jets. The first is the *muon-in-jet* correction, which consists on including the energy of the muon into the jet. Then, in 0-lepton and 1-lepton channels, a residual correction called *PtReco* is applied to recover the energy of neutrinos. In the 2-lepton channel the PtReco correction is replaced by a kinematic likelihood fit (KF) [25] that uses the full kinematic information of the event to balance the $\ell\ell bb$ system. The muon-in-jet and PtReco corrections are applied *per-jet*, while the KF is applied *per-event*. Extensive studies have been performed to optimise the muon-in-jet correction. Details are provided in Sect. 7.4.1, followed by an overview of the other $b$-jet corrections.

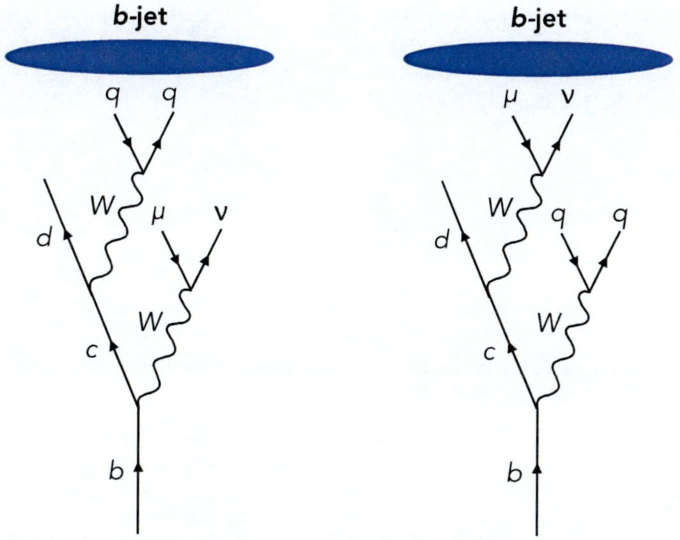


**Fig. 7.6** Examples of $b$-jet semileptonic decays at quark-level. A $b$-jet may contain at least one muon in ~21% of cases due to the decay of virtual $W$ bosons

## 7.4.1 Muon-in-jet Correction

The muon-in-jet correction, introduced in the first $VH(H \to b\bar{b})$ analysis of Run-2 [23], consists on integrating into the four-momentum of the jet the four-momentum of a muon found in the vicinity of the jet itself, while removing the four-vector of the energy deposited by the muon in the calorimeter. A reconstructed muon is considered as a candidate for the muon-in-jet correction, if it has $p_T > 5$ GeV and is found within a cone of fixed distance $\Delta R = 0.4$ with respect to the jet axis. *Medium* quality muons (as defined in 5.4.2) are considered by default. If there is more than one muon satisfying the requirements, only the closest to the jet axis is considered for the correction. For the $VH(H \to b\bar{b})$ analysis presented here, several studies on the properties of the correction muon candidates have been performed in the 0-lepton channel with the MC quark-induced signal sample $ZH \to \nu\nu b\bar{b}$, in order to identify the optimal configuration for the muon-in-jet correction.

In Fig. 7.7a, the distribution of the reconstructed muon candidate $p_T$, derived in the 0-lepton MC sample, is shown as a function of the $b$-jet $p_T$. As expected, the correction mainly concerns low $p_T$ muons (below 10 GeV) escaped from low $p_T$ jets (peaked around 30–40 GeV). As shown in Fig. 7.7b, the muon retains about 10% of the jet momentum.

These distributions have been obtained by including both the leading and sub-leading jets. However, the $p_T$ distributions are quite different between the two types of jets, with the leading one being significantly boosted, due to the 0-lepton trigger requirement of $E_T^{miss} > 150$ GeV. The $p_T$ distribution of the jets is shown in Fig. 7.8,

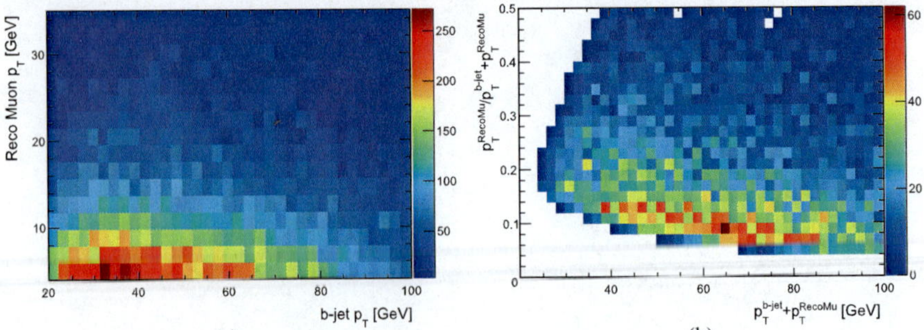


**Fig. 7.7** Distribution of the candidate muon $p_{\mathrm{T}}$ as a function of the $b$-jet $p_{\mathrm{T}}$ (**a**). About 10% of the jet momentum is carried by the muon (**b**)

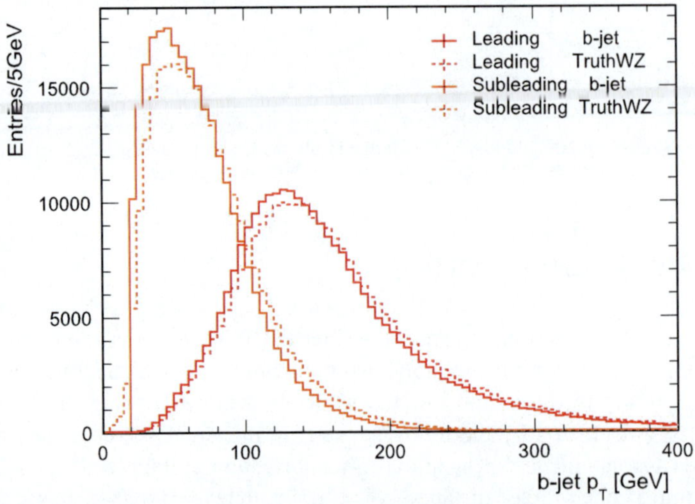


**Fig. 7.8** Distribution of the $b$-jet transverse momentum $p_{\mathrm{T}}$, shown separately for leading (red) and sub-leading (orange) jets. The distributions of the reconstructed jets (continuous line) are compared to the truth jets (dashed line)

both for the reconstructed and truth jets. The *TruthWZ* collection contains particle-level jets reconstructed with the same algorithm as for the reconstructed jet: stable particles are used, including non-isolated neutrinos and muons. *TruthWZ* jets are used as truth target for the reconstructed jets and in Fig. 7.8 are slightly boosted with respect to the reconstructed jets which do not contain muons and neutrinos yet. Different muon qualities and $p_{\mathrm{T}}$ thresholds have been compared, in order to identify the optimal selection for the candidates of the muon-in-jet correction. The comparison is based on the efficiency and fake rate to select muon candidates. These quantities are evaluated by matching the reconstructed non-isolated muon, referred to as *reco*, and a truth-level muon, indicated as *truth*, that is found to be originated from

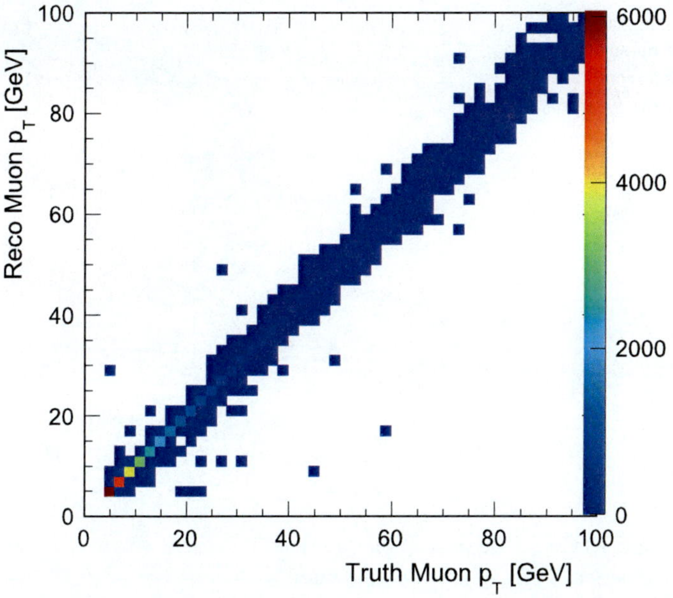


**Fig. 7.9** Transverse momentum of the reconstructed muon candidates versus the momentum of the matched truth muons that are originated from a $b$-hadron

a $b$-hadron decay. The matching requirement is $\Delta R(reco - truth) < 0.4$. Figure 7.9 shows the transverse momentum of the reconstructed muon as a function of the $p_{\mathrm{T}}$ of the *truth* muon matched. The matching performs reasonably well, as only 0.3% of matched muons is found with $|p_{\mathrm{T}}^{\mathrm{reco}} - p_{\mathrm{T}}^{\mathrm{truth}}| > 10$ GeV.

The *muon correction efficiency* is defined as the ratio between the number of matched muons and the total number of available truth $b$-initiated muons. A reconstructed muon is identified as a *fake candidate* when it is not matched to a corresponding truth muon, either because the requirement on the angular distance $\Delta R(reco - truth) < 0.4$ is not satisfied, or because a corresponding truth muon is not found within the jet. A schematic definition of muon correction efficiency and rate of fake muon candidates is shown in Fig. 7.10.

The muon correction efficiency is studied as a function of $p_{\mathrm{T}}$, starting from 4 GeV, and as a function of $\eta$ of the truth muons. The corresponding distributions are shown in Fig. 7.11a and b, for different reconstructed muon types: *loose*, *medium*, *tight* (discussed in Sect. 5.4.2). The *loose* (red) and *medium* (blue) muons provide similar results for the efficiency as a function of $p_{\mathrm{T}}$, showing a *plateau* around 95% which decreases below 90% for a tighter selection of muons (green). The rate of fake candidates is shown in Fig. 7.11c and d as a function of $p_{\mathrm{T}}$ and $\eta$ of the reconstructed muons, respectively. The smallest fake rate is provided by *tight* muons and the largest by *loose* muons. The largest fake rate (10–30%) is observed in muons with a $p_{\mathrm{T}}$

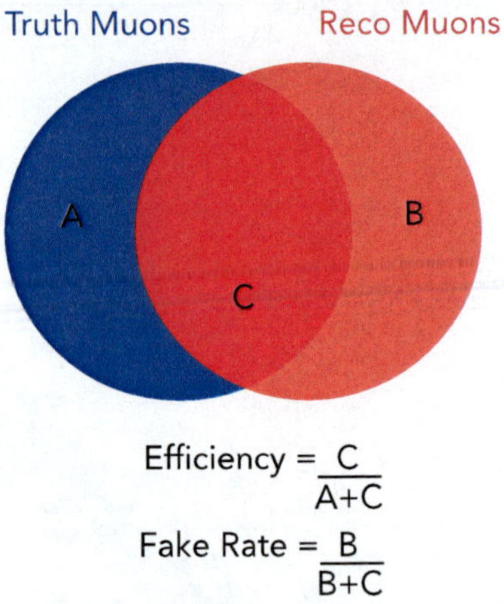


**Fig. 7.10** Schematic definition of muon correction efficiency and rate of fake candidates

between 4 and 10 GeV. The default *medium* quality for the muons provides a good compromise between efficiency and fake rate.

The multiplicity of reconstructed muons and fake candidates are evaluated per jet, for different $p_{\mathrm{T}}$ selections of *medium* muons. With a muon selection at $p_{\mathrm{T}} > 4$ GeV, 13% of jets contain one muon of which 9% are identified as fake candidates. A total of 0.5% of jets contains two muons, of which at least one is fake in 26% of the cases. By increasing the $p_{\mathrm{T}}$ cut of the muon selection, the number of jets containing muons decreases, as well as the amount of corresponding fake candidates. A study on the origin of fake muon candidates reveals that, in case of only one muon considered within the jet, more than 68% of fake candidates are caused by a wrong muon identification due to detector mis-measurements at low $p_{\mathrm{T}}$ (*background muons*), 4% are non-isolated muons and for the remaining 28% the origin is unknown. If a second muon candidate within the jet is found to be fake, this is a *background muon* in about 79% of cases, and non-isolated in 16%. The good muon candidates (not fake) are purely non-isolated muons, if only one candidate is considered. If a second candidate is available, it originates from *background muons* in 10% cases. This suggests not to include the second muon into the jet since the rate of fake candidates would increase and the good muon candidates would be contaminated by background.

Since fake candidates mostly have very low $p_{\mathrm{T}}$, an increment on the $p_{\mathrm{T}}$ cut from 4 GeV to 10 GeV would remove about 80% of fake muons, corresponding to 67% of the total momentum carried by fake muons, and wrongly used for correcting the energy of the jets. Nevertheless, about 40% of good muon candidates would be removed with the 14% of the total momentum of good muons, reversing the benefit of the muon-in-jet correction.

Ultimately, the most appropriate figure of merit to evaluate the best muon selection for the muon-in-jet correction is the resolution of the $m_{bb}$ distribution. The invariant mass $m_{bb}$ for different $p_{\mathrm{T}}$ cuts (with fixed *medium* type) and different muon types (with fixed $p_{\mathrm{T}}$ cut at 4 GeV) is evaluated on the $ZH \to \nu\nu b\bar{b}$ MC sample and

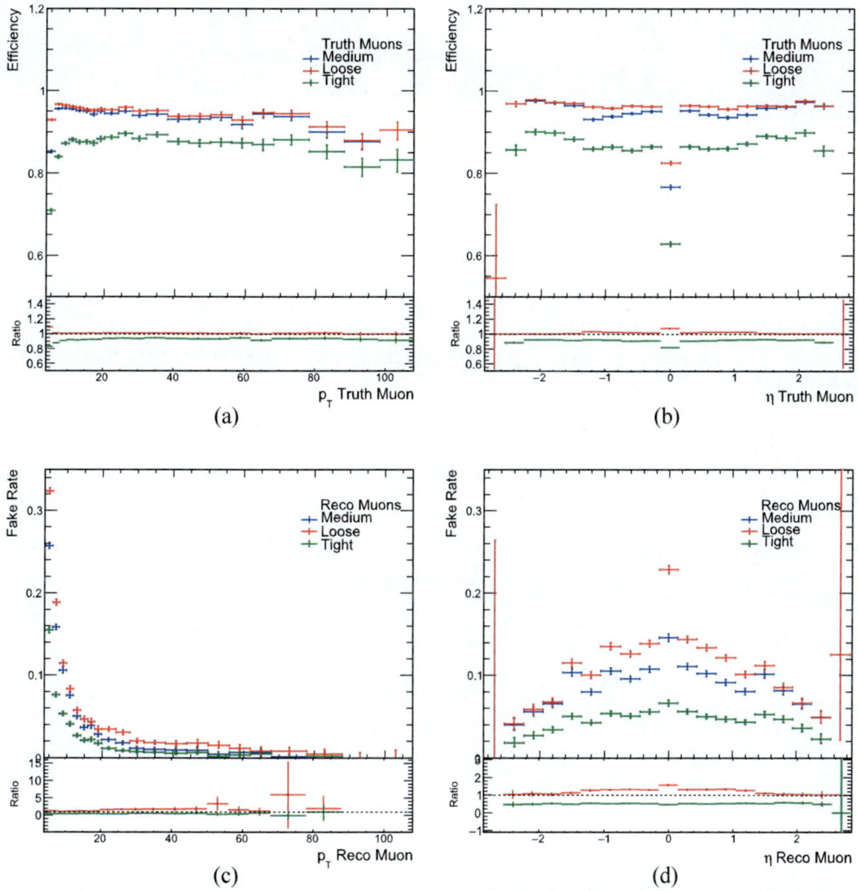


**Fig. 7.11** Top row: efficiency of the muon-in-jet candidates as a function of the truth muon $p_T$ (**a**) and as a function of $\eta$ (**b**), for various muon types. Bottom row: fake rate of the muon-in-jet candidates as a function of the reco muon $p_T$ (**c**) and as a function of $\eta$ (**d**), for various muon types. The lower insert of the plots represents the ratio with respect to the efficiency or fake rate corresponding to the *medium* selection

shown in Fig. 7.12a and b, respectively. A Bukin function [26] is used to fit the $m_{bb}$ distribution and extract the values of the mean and the width, which are reported in Table 7.4a and b.

The values extracted with the fit show that the narrower distribution is provided by selecting *medium* muons with $p_T > 4$ GeV, thus identifying this as an optimal working point. However, due to a stringent analysis timescale, a nearly optimal selection of *medium* muons with $p_T > 5$ GeV was used for the analysis presented here. Systematic uncertainties related to the muon selection for the muon-in-jet correction have been studied in Run-1 [23] and found to be negligible.

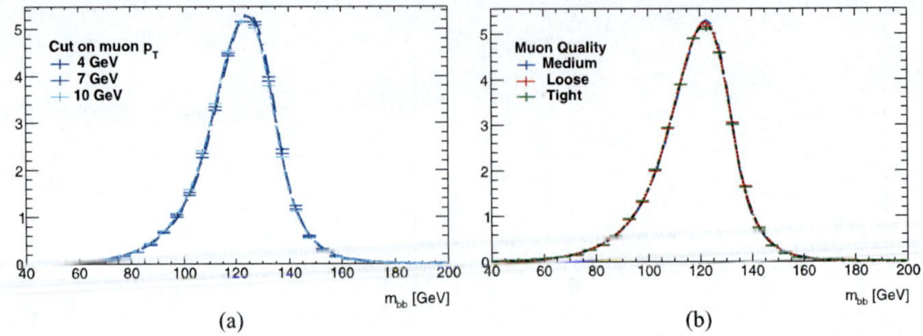


**Fig. 7.12** Invariant mass $m_{bb}$ distributions for different $p_{\mathrm{T}}$ cuts (**a**) and muon types (**b**), evaluated with the $ZH \to \nu\nu b\bar{b}$ sample

**Table 7.4** Mean and width obtained with a Bukin fit of the $m_{bb}$ distributions. Results in (a) refer to muons selected with *medium* quality and different $p_{\mathrm{T}}$ cuts; results in (b) refer to muons selected with $p_{\mathrm{T}} > 4$ GeV and different qualities

| | Mean | Width |
|---|---|---|
| (a) | | |
| 4 GeV | $122.4 \pm 0.1$ | $11.39 \pm 0.06$ |
| 7 GeV | $121.9 \pm 0.1$ | $11.56 \pm 0.05$ |
| 10 GeV | $121.8 \pm 0.1$ | $11.65 \pm 0.05$ |
| (b) | | |
| *medium* | $122.4 \pm 0.1$ | $11.39 \pm 0.06$ |
| *loose* | $122.0 \pm 0.1$ | $11.51 \pm 0.05$ |
| *tight* | $121.9 \pm 0.1$ | $11.59 \pm 0.05$ |

### 7.4.2 PtReco Correction

The PtReco correction is derived from simulation and applied to the 0-lepton and 1-lepton channels as a function of $p_{\mathrm{T}}$, to further improve the jet response and account for undetected neutrinos. It is based on the residual difference in the jet response between the reconstructed $b$-jets, which had received the muon-in-jet correction in addition to the standard corrections, and the corresponding truth jets which include muons and neutrinos (*TruthWZ*). Two categories of jets are considered, depending on the decay of the $b$-hadron inside the jet. If at least one muon is reconstructed inside the jet, the jet is considered *semileptonic*, otherwise the decay is classified as *hadronic*. The correction factors are derived in a simulated $ZH \to \ell\ell b\bar{b}$ sample in bins of $p_{\mathrm{T}}$ of the reconstructed jet, separately for the two categories of jets. For semileptonic jets, the jet four-vector, already corrected with the muon contribution, receives an increment of about 25% for jets with low $p_{\mathrm{T}}$. The correction decreases as the transverse momentum increases, reaching a plateau of 6% at $p_{\mathrm{T}} \approx 100$ GeV.

The PtReco for the hadronic jets corrects the momentum of about 12% at low $p_T$ and reaches a plateau of 1% at $p_T \approx 70$ GeV [21].

### 7.4.3 Kinematic Fit

In the 2-lepton channel, the $ZH \rightarrow \ell\ell b\bar{b}$ event kinematics is fully reconstructed. The hadronic energy is balanced with the leptonic contribution, arising from the $Z$ boson decay, which is measured with high precision. A per-event kinematic likelihood fit is applied to exploit the kinematic balance and improve the estimate of the $b$-jet energy. The likelihood function is parametrised with a combination of functions: Gaussian distributions for the transverse momenta $p_T$ of the two leptons, as well as the third jet, when available; Gaussian of the transverse momentum of the $\ell\ell bb$ system; transfer functions relating the truth $p_T$ to the reconstructed $p_T$ of the two $b$-jets; Breit–Wigner constraint on the dilepton mass $m_{\ell\ell}$.

Figure 7.13 shows the invariant mass $m_{bb}$ as additional corrections are applied to the jet energy. The result shown is obtained with 36.1 fb$^{-1}$ dataset of the previous analysis iteration, as the configuration of the jet corrections has not changed since then. The distribution obtained with the inclusion of the muon-in-jet correction (blue curve) shows an improved resolution of 15% with respect to the result obtained with the only standard jet calibration (black curve) in the 2-lepton channel. The

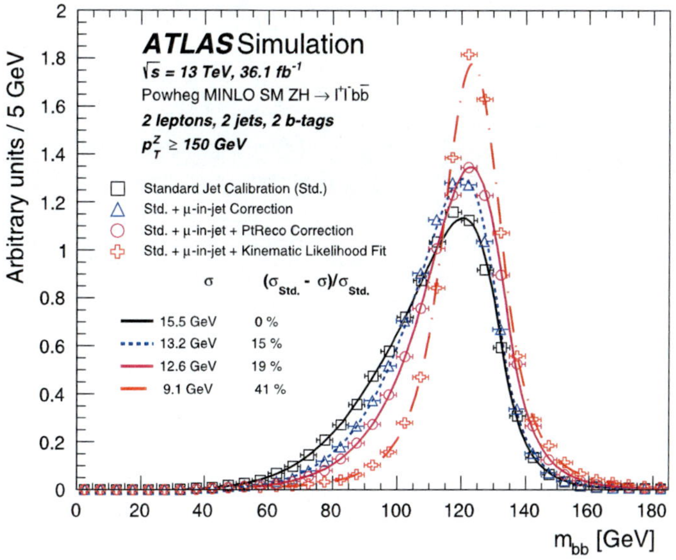


**Fig. 7.13** Comparison in 2-lepton channel of the $m_{bb}$ distribution as additional energy corrections are included in the jet correction. *Image source* [23]

PtReco correction augments the $m_{bb}$ resolution of 19% (purple), while adding the KF correction provides an overall improvement of 40% (red).

## 7.5 Multivariate Analysis

The main strategy of the $VH(H \to b\bar{b})$ analysis is based on an MVA approach, introduced in Sect. 3.2.1. In order to maximise the discrimination between signal and background events passing the selection (Sect. 7.3), a BDT is trained separately for each lepton channel and for each signal region. A set of observables are combined into each BDT, designed to separate the Higgs boson events from expected backgrounds, yielding the final discriminant variable used as input of the statistical analysis. The set of observable changes across the channels, due to the different background compositions. A summary of the variables used is listed in Table 7.5. The two $b$-tagged jets are indicated by $b_{1(2)}$, $V$ is the reconstructed vector boson, $m_{\text{eff}}$ represents the scalar sum of $E_{\text{T}}^{\text{miss}}$ and the $p_{\text{T}}$ of all signal jets present in the event. $S_{\text{T}}$ is the scalar sum of the $p_{\text{T}}$ of the leptons and jets in the event and $E_{\text{T}}^{\text{miss}}/\sqrt{S_{\text{T}}}$ is

**Table 7.5** Variables used for the multivariate discriminant in each of the categories. *Table source* [8]

| Variable | 0-lepton | 1-lepton | 2-lepton |
|---|---|---|---|
| $p_{\text{T}}^{V}$ | $\equiv E_{\text{T}}^{\text{miss}}$ | × | × |
| $E_{\text{T}}^{\text{miss}}$ | × | × | |
| $p_{\text{T}}^{b_1}$ | × | × | × |
| $p_{\text{T}}^{b_2}$ | × | × | × |
| $m_{bb}$ | × | × | × |
| $\Delta R(\vec{b_1}, \vec{b_2})$ | × | × | × |
| $\lvert\Delta\eta(\vec{b_1}, \vec{b_2})\rvert$ | × | | |
| $\Delta\phi(\vec{V}, \vec{bb})$ | × | × | × |
| $\lvert\Delta\eta(\vec{V}, \vec{bb})\rvert$ | | | × |
| $m_{\text{eff}}$ | × | | |
| $\min[\Delta\phi(\vec{\ell}, \vec{b})]$ | | × | |
| $m_{\text{T}}^{W}$ | | × | |
| $m_{\ell\ell}$ | | | × |
| $E_{\text{T}}^{\text{miss}}/\sqrt{S_{\text{T}}}$ | | | × |
| $m_{\text{top}}$ | | × | |
| $\lvert\Delta Y(\vec{V}, \vec{bb})\rvert$ | | × | |
| | Only in 3-jet events | | |
| $p_{\text{T}}^{\text{jet}_3}$ | × | × | × |
| $m_{bbj}$ | × | × | × |

**Table 7.6** Configuration of the hyper-parameters used for the TMVA implementation of the BDT applied in the $VH(H \rightarrow b\bar{b})$ analysis

| TMVA Setting | Value | Description |
|---|---|---|
| BoostType | AdaBoost | Boost procedure |
| AdaBoostBeta | 0.15 | Learning rate |
| SeparationType | GiniIndex | Node separation gain |
| PruneMethod | NoPruning | Pruning method |
| NTrees | 200 | Number of trees |
| MaxDepth | 4 | Maximum tree depth |
| nCuts | 100 | Number of cuts per variable and node |
| nEventsMin | 5% | Minimum number of events in a node |

defined as the $E_{\mathrm{T}}^{\mathrm{miss}}$ *significance*. Finally, $|\Delta Y(\vec{V}, \vec{bb})|$ is the difference in rapidity between the Higgs boson candidate and the $W$ boson candidate. The most powerful variables for the discrimination are the invariant mass of the dijet system $m_{bb}$, the angular distance between the two jets $\Delta R_{bb}$ and the transverse momentum of the vector boson $p_{\mathrm{T}}^{V}$. The BDT is implemented with the ROOT-based TMVA (Toolkit for Multivariate Data Analysis) package [27]. Since the multivariate algorithms must be trained and evaluated on independent samples in order to ensure a robust and unbiased result, the MC events are split into training and validation samples. A *k-fold* cross-validation resampling procedure [28] is applied: this is a common procedure in ML applications that consists in splitting the training set into $k$ smaller sets of approximately equal size. For each of the $k$ folds the model is trained using $k-1$ folds as training data and the remaining part of the data are used to validate the model itself. In our implementation a *2-fold* approach is used, in such a way that the dataset is split into two sub-sets based on the event number, either *odd* or *even*, each composed by 50% of the total events. Each of the two orthogonal subsets is used as training and validation set in two consecutive steps. The results of the two training stages, performed on two different sub-sets, are expected to be unbiased. The final discriminant is then constructed by the combination of the two resulting discriminants. The set of training parameters (or *hyper-parameters* of the discriminating model) used for the BDT is shown in Table 7.6.

### *7.5.1 MVA Hyper-Parameters Studies*

The set of training parameters was optimised for the Run-1 analysis [23] through a one-dimensional scan of the parameters. Preliminary studies for a similar optimisation have been performed with the aim to select the most suitable hyper-parameters setup for the 0-lepton channel in the current analysis. Several configurations have

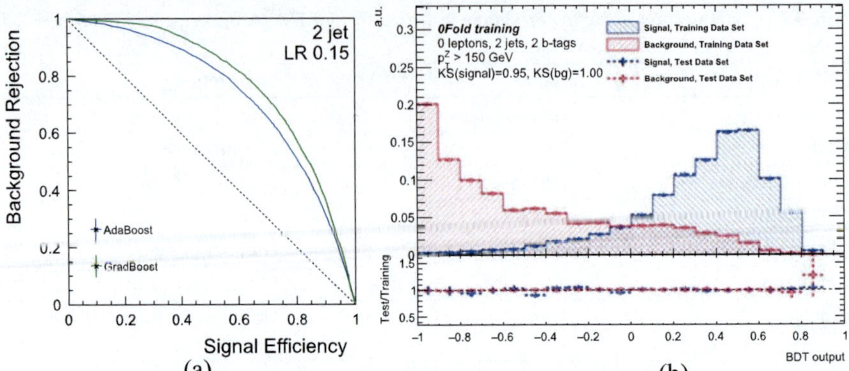


(a) (b)

**Fig. 7.14** ROC comparison between AdaBoost and GradBoost performances in 2-jet (**a**) and distribution of the BDT discriminant obtained with the GradBoost algorithm (**b**). The learning rate is fixed at the default value (LR = 0.15), as well as the other parameters

been obtained by changing one parameter at a time and retraining the BDT model. The performance of a model is evaluated on the basis of the *Receiver Operating Characteristic* (ROC). The ROC curve illustrates the performance of a binary classifier system as a variable threshold is applied on the discriminant score. The curve is created by plotting the signal acceptance against the background rejection at various threshold settings. One of the most common estimators to evaluate the ROC performance is its *Area Under the Curve* (AUC). A unitary value of the ROC AUC indicates a perfect discrimination of the signal from background, while a value of 0.5 implies no discrimination ability of the model.

**Boost Type**

As described in Sect. 3.2.1, different boosting algorithms can be used to train a BDT model. The GradBoost [29] algorithm is used to retrain the model and the performance is compared with the result obtained with AdaBoost, used in the default training. In Fig. 7.14a the corresponding ROC curves obtained with the testing samples are compared in the 2-jet category, showing a significant improvement when GradBoost is used. In terms of AUC, the improvement corresponds to 5% for the 2-jet category and 4% for the 3-jet category. In Fig. 7.14b the distribution of the BDT discriminant obtained with the GradBoost implementation in 2-jet category is shown as an example.

**Training Statistics**

The *2-fold* technique employed in the analysis determines the training statistics, since it divides the dataset into two equal sub-sets. However, the *k-fold* technique allows to use an arbitrary value for $k$, thus allowing to change the proportion between training and testing samples. As the number of folds increases, a larger training sample can be used while the validation sample is reduced. This, in principle, would improve the accuracy of the BDT model. However, the testing sample must be large enough to

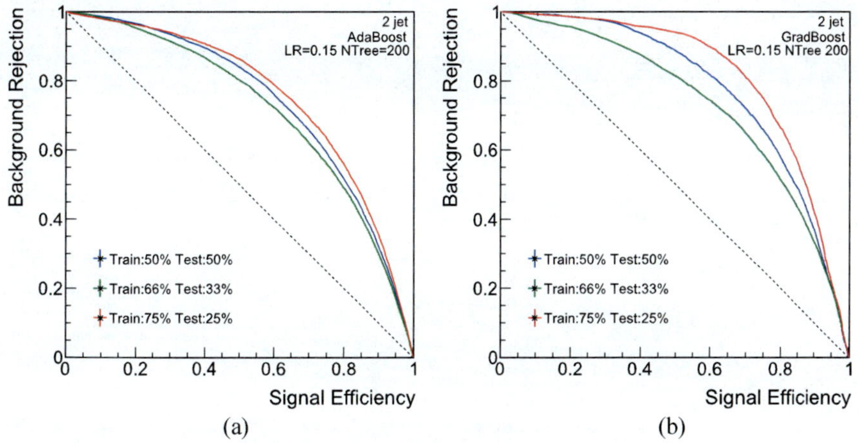


**Fig. 7.15** ROC comparison between different training statistics used with AdaBoost (**a**) and GradBoost (**b**). The blue curve is obtained with a *2-fold* scheme, the green with *3-fold* and the red curve with *4-fold*. The results are shown only for the 2-jet category

be statistically representative of the whole dataset [30]. A large $k$ also means that the model needs to be trained several times and that could be computationally expensive. After due consideration, different $k$ values are used to train the BDT model and the results compared in Fig. 7.15 both for AdaBoost and GradBoost.

**Learning Rate**

The learning rate is an important hyper-parameter that, as anticipated in Sect. 3.2.1, controls the adjustment of the weights of the BDT at each boosting stage. The smaller is the learning rate, the slower is the convergence to the minimum of the objective function. With a small learning rate, the model requires more trees to build a robust BDT, but the accuracy of its prediction is expected to improve [31]. On the other hand, with a large learning rate, the global minimum of the objective function may be overshot and the model may not converge. Different values of learning rate have been tested both with AdaBoost and GradBoost algorithm, while keeping the other hyper-parameters fixed to the default values. The values of learning rate tested are the following: 0.001, 0.01, 0.15 (default), 0.3, 0.5. The ROC comparison for GradBoost is shown in Fig. 7.16a for the 2-jet category. Very small values for the learning rate, such as 0.001 (blue curve) and 0.01 (green curve), make the fit model not converging and the BDT loses the ability to discriminate between signal and background. The corresponding BDT distribution (obtained with the lowest learning rate tested) looks as shown in Fig. 7.16b. The default value for the learning rate 0.15 (red curve) provides the best performance among the different values tested.

**Number of Trees**

Another parameter that has been explored is the number of trees in the BDT model. Different models have been constructed with number of trees equal to 200 (default), 400, 800, 1000 and 2000. The best performance in terms of ROC AUC is provided

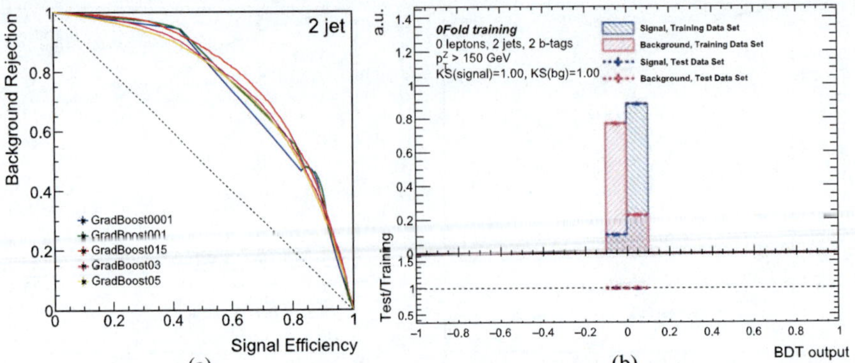


(a) (b)

**Fig. 7.16** ROC comparison between different learning rates used with GradBoost, in 2-jet **a** category. In Figure **b** the BDT distribution obtained with a learning rate of 0.001 is shown

by the default value, with small deterioration for models with a larger number of trees, both in the 2-jet and 3-jet category.

In summary, the studies on the MVA hyper-parameters show that a significant improvement can be achieved with the GradBoost algorithm. Further investigations are needed to draw conclusions on the other parameters evaluated. It is important to notice that most of the hyper-prameters are correlated and more detailed studies should take these correlations into account. A possible way to proceed with the hyper-prameters tuning to find an optimal configuration would be a grid search, which consists on exhaustively testing the parameter combinations within a given parameter space.

## 7.5.2 *BDT Transformation*

The default binning of the output distributions resulting from the multivariate analysis is not optimal and must be transformed to maximise the final BDT sensitivity. An iterative algorithm is applied to remap the BDT histogram according to the transformation Z, defined as:

$$Z = z_s \frac{n_s}{N_s} + z_b \frac{n_b}{N_b}, \tag{7.1}$$

where $N_{s(b)}$ is the total number of signal (background) events in the BDT histogram, $n_{s(b)}$ is the number of events in a predefined interval within the BDT output range and $z_{s(b)}$ are free parameters used to tune the algorithm and need to be optimised. In this analysis, optimal values are found to be $z_s = 10$ and $z_b = 5$. This requirement guarantees that the MC statistical uncertainty is kept below 20% in each bin. The iterative algorithm starts by grouping the bins corresponding to the highest values of the discriminant (the last bins on the right). The adjacent bins on the left are progressively incorporated one by one and the $Z$ function is recomputed at every iteration. When the condition $Z > 1$ is reached, a single bin is constructed from the grouped bins and the procedure is repeated starting from the next bin on the left. The procedure is repeated until all the BDT distribution is completely re-binned such that across the bins the statistical uncertainty is uniform.

## 7.6 Statistical Analysis

After training the BDT on each SR, a statistical fitting procedure is employed to extract the strength of the Higgs boson signal from the data. The signal strength $\mu$ is defined as the observed product of the Higgs boson production cross section and its branching ratio into $b\bar{b}$, in units of the corresponding SM values:

$$\mu = \frac{\sigma \times \mathrm{BR}}{\sigma_{\mathrm{SM}} \times \mathrm{BR}_{\mathrm{SM}}}. \tag{7.2}$$

The statistical procedure that extracts the signal strength is based on a binned maximum likelihood fit, simultaneously performed in the three channels. The likelihood function $\mathcal{L}(\mu)$ is built as the product of the Poisson probability terms over the $N$ bins of the input distributions:

$$\mathcal{L}(\mu) = \prod_{i=1}^{N} \frac{(\mu s_i + b_i)^{n_i} e^{-(\mu s_i + b_i)}}{n_i!} \tag{7.3}$$

In Eq. (7.3), $s_i$ and $b_i$ are the expected signal and background yields in the $i^{th}$ bin, respectively, and $n_i$ is the number of observed data. The inputs to the likelihood function include the unidimensional re-binned BDT discriminants in all the categories, for a total of eight SRs. In addition, for the two $W$+HF CRs the event yields are used as input to the fit, while the $m_{bb}$ distributions are employed for the four $e\mu$ CRs.

The principal parameter of interest that must be extracted in the fit is the signal strength $\mu$. The effect of the systematic uncertainties is embodied in the likelihood as well, with the introduction of nuisance parameters (NP) $\boldsymbol{\theta}$. The prior knowledge of each NP enters the fit model as a Gaussian distribution, thus modifying the likelihood function, which becomes a function of $\mu$ and $\boldsymbol{\theta}$:

$$\mathcal{L}(\mu, \boldsymbol{\theta}) = \prod_{i=1}^{N} \frac{(\mu s_i(\theta) + b_i(\theta))^{n_i} e^{-(\mu s_i(\theta) + b_i(\theta))}}{n_i!} \times \prod_{j=1}^{N_\theta} \frac{1}{\sqrt{2\pi}\sigma_j} e^{-\frac{(\theta_j - \theta_{0j})^2}{2\sigma_j^2}}, \tag{7.4}$$

where $\theta_j$ is the NP to be fitted, $\theta_{0j}$ and $\sigma_j$ are the central value and the uncertainty assigned to the prior, respectively. The inclusion of the NPs in the likelihood adds flexibility to the model while reducing the analysis sensitivity. In order to test a certain hypothesis for the value of $\mu$, a profile likelihood ratio is considered as a test statistic [32]:

$$\lambda(\mu) = \frac{\mathcal{L}(\mu, \hat{\hat{\theta}})}{\mathcal{L}(\hat{\mu}, \hat{\theta})}, \tag{7.5}$$

where $\hat{\mu}$ and $\hat{\theta}$ are the parameters that maximise the likelihood, while $\hat{\hat{\theta}}$ denotes the value of $\theta$ that maximises the likelihood for a given $\mu$. The likelihood ratio is defined in the range [0, 1], describing good agreement between the data and the specific hypothesis of $\mu$ when $\lambda(\mu)$ is near 1, and disagreement as $\lambda(\mu)$ moves to the null value. It is convenient to construct the test statistic $t_\mu$:

$$t_\mu = -2 \ln \lambda(\mu). \tag{7.6}$$

As the value of $t_\mu$ gets larger, the incompatibility between the data and the given hypothesis $\mu$ increases. The disagreement can be quantified by the *p-value* computed from the test statistic:

$$p_\mu = \int_{t_{\mu,obs}}^{\infty} f(t_\mu|\mu)\, dt_\mu, \tag{7.7}$$

where $t_{\mu,obs}$ is the observed value of $t_\mu$ in data and $f(t_\mu|\mu)$ describes the probability density functions of $t_\mu$ under the assumption $\mu$. The relation between $t_\mu$ and p-value is shown in Fig. 7.17a. The p-value can be interpreted as the probability, under the hypothesis $\mu$, of finding data that are incompatible with the hypothesis. The p-value is related to the *significance* $Z$, defined as

$$Z = \Phi^{-1}(1 - p_\mu), \tag{7.8}$$

where $\Phi^{-1}$ is the quantile (inverse of the cumulative distribution) of a standard Gaussian. The relation of the p-value and the significance $Z$ is shown in Fig. 7.17b. A p-value of $2.87 \times 10^{-7}$ corresponds to a significance $Z = 5$, which represents a deviation of $5\sigma$ with respect to the background-only hypothesis. The physics community considers this value of significance as a sufficient level to reject the background-only hypothesis and claim an *observation* or *discovery*.

Assuming that a new signal can only increase the mean event rate with respect to the case of background-only, the signal process must have $\mu \geq 0$. For the special case of testing the $\mu = 0$ hypothesis, the test statistic $t_\mu$, defined in Eq. (7.6), is modified as follows:

$$q_0 = \begin{cases} -2 \ln \lambda(0) & \hat{\mu} \geq 0, \\ 0 & \hat{\mu} < 0, \end{cases} \tag{7.9}$$

where $\lambda(0)$ is the profile likelihood ratio for $\mu = 0$. With the given assumptions, the $\mu = 0$ hypothesis is effectively rejected, leading to the discovery of new signal, only when $\mu \geq 0$. Measuring an increased yield with respect to the expected background

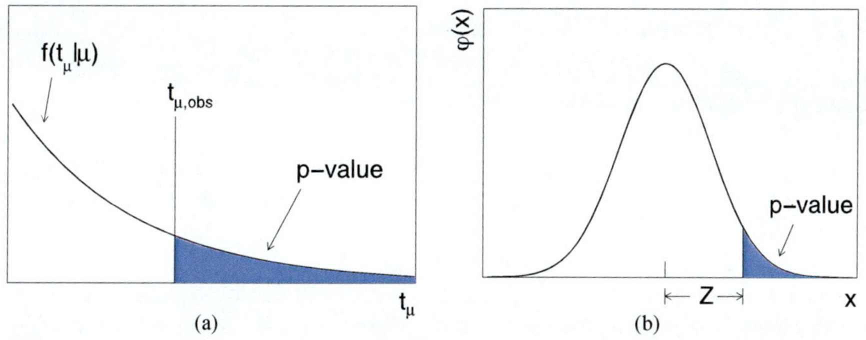


**Fig. 7.17** Relation between the p-value and the test statistic $t_\mu$ (**a**). Relation between the p-value and the significance $Z$ (**b**). *Image source* [32] (use permitted under the Creative Commons Attribution Noncommercial License CC BY-NC 2.0)

($\mu > 0$) corresponds to find a larger value of $q_0$. As $q_0$ gets larger, the incompatibility of data with the $\mu = 0$ hypothesis increases.

In the $VH(H \rightarrow b\bar{b})$ analysis the results are presented in terms of probability that the background-only ($\mu = 0$) hypothesis is compatible with the data. The fitted $\hat{\mu}$ value is obtained by maximising the likelihood function with respect to all parameters. The profile likelihood function assumes a parabolic shape around the minimum. The uncertainty on $\mu$ is obtained with a profile likelihood method by varying $q_0 = -2 \ln \lambda(\mu)$ of one unit, corresponding to a variation of $\pm 1\sigma$ on $\mu$. Expected results are obtained by replacing the data with a representative dataset, the *Asimov* dataset, constructed from simulations by setting all the NPs to their best-fit values obtained with data. Due to the finite number of events contained in the MC samples used in the analysis, the background estimates are subject to statistical fluctuations that must be considered in the fit. This is done with the Beeston–Barlow technique [33], which models the number of events in a bin as a Poisson distribution and treats the mean distribution as additional NP in the likelihood model.

## 7.7 Systematic Uncertainties

The $VH(H \rightarrow b\bar{b})$ analysis is affected by several systematic uncertainties, which can be grouped into different categories, according to their origin: experimental uncertainties, simulated signal and EW background uncertainties, data-driven multi-jet background uncertainties.

### 7.7.1 Experimental Uncertainties

The experimental uncertainties depend on the process of reconstruction and correction of the objects used in the analysis, as well as on the detector performances. The dominant sources are found to be related to the jet flavour-tagging and the jet energy calibration. The first type is derived from the difference in flavour-tagging efficiency between data and simulation, measured as scale factors. An event weight systematic is retrieved separately for $b$-jets, $c$-jets and light-jets, which have uncertainties estimated from multiple measurements. These uncertainties are separated in several uncorrelated components that are treated independently. The uncertainty due to the tagging efficiency is around 2% for $b$-jets, 10% for $c$-jets and 30% for light-jets.

A separate uncertainty is evaluated for the extrapolation of the $b$-jet efficiency calibration for $p_T > 300$ GeV and the mis-identification of hadronically decaying $\tau$-leptons as $b$-jets. Several uncertainties are considered for the JES and JER corrections, which are derived as described in Sect. 5.5.2. A set of 23 uncorrelated parameters that take into account the single steps of the jet energy calibration and resolution is used to evaluate this systematic uncertainty.

The systematic uncertainties due to the presence of leptons in the analysis are evaluated using the scale factors obtained for their reconstruction, identification and isolation. The uncertainties are evaluated through variations of the scale factors and they are found to have a small impact on the final measurements. The lepton and jets systematic uncertainties are combined together to evaluate the uncertainties on the $E_T^{miss}$, to which contributions due to the scale, resolution and efficiency of the $E_T^{miss}$ soft term, defined in Sect. 5.7, are added.

The systematic uncertainty arising from the integrated luminosity is evaluated by combining the scale calibration results of 2015, 2016 and 2017 obtained with van der Meer scans [34]. The resulting uncertainty is 2%. The pile-up also contributes to the set of uncertainties. A correction factor of 1.03 is applied to the nominal value of the number of interactions per bunch crossing and is varied to 1.00 and 1.18 to estimate the up and down uncertainties.

### 7.7.2 Simulated Sample Uncertainties

The assumptions made in the simulation of the signal and background processes originate additional systematic uncertainties. In general, these uncertainties can affect the normalisation, the shape and the event acceptance of the kinematic distributions involved in the measurement. The normalisation of the main backgrounds $t\bar{t}$, $W$+HF and $Z$+HF are left unconstrained (floating) and directly extracted from data with the global likelihood fit. For the remaining simulated background processes, the overall normalisations and associated uncertainties are obtained from very accurate cross section calculations. Since these uncertainties are due to MC assumptions, they are usually assessed through comparisons at particle-level between the nominal simulated sample and an alternative choice. The alternative choice is either produced by a different generator or by varying the nominal values of some parameters of the generator used in the simulation. The nominal and alternative samples are normalised using the same production cross section. For each variation, acceptance and shape uncertainties are evaluated. The first type accounts for the variations in the number of events falling in a certain analysis category and the possible migration from a category A to a category B. The acceptance uncertainties are calculated as *double-ratios* between acceptances $\mathcal{A}$, as:

$$\frac{\mathcal{A}(\text{Category}_A^{nom})}{\mathcal{A}(\text{Category}_B^{nom})} \Bigg/ \frac{\mathcal{A}(\text{Category}_A^{alt})}{\mathcal{A}(\text{Category}_B^{alt})}, \tag{7.10}$$

where each ratio is obtained between the number of events falling in the two categories, the first calculated with the nominal MC sample (*nom*) and the other using an alternative one (*alt*). Shape uncertainties are addressed separately for each analysis region and derived for the $m_{bb}$ and $p_{\mathrm{T}}^{V}$ variables as the variation induced in these distributions by an alternative generator.

The simulated sample uncertainties are evaluated for the signal and for each of the background samples separately, as described in the following.

**$VH(H \to b\bar{b})$ signal** The signal samples are normalised to the best theoretical prediction of the cross section for the different processes. The systematic uncertainties in the overall $VH$ production cross section are obtained by varying the *renormalisation* and *factorisation* scale factors, as well as the PDF sets. Acceptance and shape uncertainties originate from missing higher-order QCD and EW corrections, PDF+$\alpha_S$ variations, differences for parton shower and underlying event models.

**$V$+jets** The normalisation of the $V$+HF is derived from the fit, separately for the 2-jet and 3-jet categories. In the case of $W$+HF background, a dedicated acceptance uncertainty has been evaluated to account for the correlation between the 1-lepton SR and the $W$+HF CR, as well as between 0-lepton and 1-lepton regions. Similarly, for the $Z$+HF background, the acceptance uncertainty was considered between the 0-lepton and 2-lepton regions. For the remaining $V + cl$ and $V + ll$, which are suppressed by the two $b$-jets requirement of the analysis and that account for less than 1% of the background, only the normalisation uncertainty is considered. For the $p_{\mathrm{T}}^{V}$ and $m_{bb}$ distributions, shape uncertainties are derived from the comparison between the nominal and alternative MC generators or variations of internal generator parameters and is parametrised as a straight line.

**Top pair** The $t\bar{t}$ background contributes to all the analysis channels and its effect on the systematic uncertainties is evaluated independently for the $0 + 1$ lepton channels and the 2-lepton channel. The 2-lepton channel involves two different kinematic regions: 75 GeV $< p_{\mathrm{T}}^{V} <$ 150 GeV and $p_{\mathrm{T}}^{V} >$ 150 GeV. For the 2-lepton channel, the normalisations are left floating, both in the 2- and 3-jet categories, and extracted from the corresponding $e\mu$ CRs. For the $0 + 1$ lepton channels, acceptance variations are considered between 2-jet and 3-jet categories, between $W$+HF control and signal regions, and between 1-lepton and 0-lepton channels. Shape uncertainties are estimated through the comparison between the nominal and alternative samples for the relevant kinematic distributions, separately in the $0 + 1$ lepton and 2-lepton channels.

**Single top** The single top contribution arises from different production channels: $Wt$, $t$- and $s$-channels. Normalisation, acceptance and shape uncertainties are evaluated for the first two, while only the normalisation uncertainty is considered for the $s$-channel, whose contribution is negligible. The normalisation uncertainties account for renormalisation and factorisation scale variations, the $\alpha_S$ uncertainty and the errors on the PDFs.

**Diboson** Three different contributions are considered: $WZ$, $WW$ and $ZZ$. For the $WW$, whose contribution is very small (it accounts for less than 0.1% of the total background), only a normalisation uncertainty is evaluated. For the main contributions, $WZ$ and $ZZ$, normalisation, acceptance and shape uncertainties are assigned, taking into account event migrations between 0-lepton and 1-lepton and between 2-jet and 3-jet categories.

### 7.7.3 *Multi-jet Background Uncertainties*

As described in Sect. 7.3.3, after the event selection, the multi-jet background is not negligible only in the 1-lepton channel, where it is estimated from data. Systematic uncertainties can affect the shape of the $m_T^W$ distributions used in the multi-jet template fits, thus affecting the background normalisation, or even the BDT distributions used as input of the final fit. Normalisation and shape uncertainties are evaluated throughout variations of the multi-jet selection: by requiring two $b$-tagged jets instead of one, by tightening the isolation requirements, by modifying the single-lepton trigger. Moreover, contaminations from $V$+jets and top quark processes are considered.

## 7.8 Results

The results of the $VH(H \to b\bar{b})$ search performed with data collected between 2015 and 2017 by the ATLAS detector, corresponding to an integrated luminosity of 79.8 $fb^{-1}$ at a centre of mass energy $\sqrt{s} = 13$ TeV, are reported in this section. The result of the global likelihood fit indicates that, in the combination of the three lepton channels, for a Higgs boson mass of 125 GeV, the probability $p_0$ of obtaining a signal as strong as the observation, by assuming the background-only hypothesis is $5.3 \cdot 10^{-7}$, while the expected value is $7.3 \cdot 10^{-6}$. The value of the signal strength parameter is:

**Table 7.7** Measured signal strengths, expected and observed p-values $p_0$ and significance values (in standard deviations) from the combined fit. *Image source* [8]

| Signal strength | Signal strength | $p_0$ | | Significance | |
|---|---|---|---|---|---|
| | | Exp. | Obs. | Exp. | Obs. |
| 0-lepton | $1.04^{+0.34}_{-0.32}$ | $9.5 \cdot 10^{-4}$ | $5.1 \cdot 10^{-4}$ | 3.1 | 3.3 |
| 1-lepton | $1.09^{+0.46}_{-0.42}$ | $8.7 \cdot 10^{-3}$ | $4.9 \cdot 10^{-3}$ | 2.4 | 2.6 |
| 2-lepton | $1.38^{+0.46}_{-0.42}$ | $4.0 \cdot 10^{-3}$ | $3.3 \cdot 10^{-4}$ | 2.6 | 3.4 |
| $VH(H \to b\bar{b})$ combination | $1.16^{+0.27}_{-0.25}$ | $7.3 \cdot 10^{-6}$ | $5.3 \cdot 10^{-7}$ | 4.3 | 4.9 |

$$\mu^{bb}_{VH} = 1.16 \pm 0.16(\text{stat.})^{+0.21}_{-0.19}(\text{syst.}) = 1.16^{+0.27}_{-0.25}.$$

An excess on the difference between the data and the fitted background is observed with a significance of 4.9 standard deviations, to be compared to an expectation of 4.3 standard deviations. These results are summarised in Table 7.7, together with the results obtained for each lepton channel separately.

Figure 7.18 shows the BDT output distributions, after the global likelihood fit, in the most sensitive high-$p_T^V$ regions of 0-lepton (top row), 1-lepton (middle row) and 2-lepton (bottom row). Shown as stacked filled histograms, the background predictions have been obtained by normalising the backgrounds and setting the systematic uncertainties according to the values of the floating normalisation and NPs extracted by the fit. The signal sample (red histogram on top of the backgrounds) is normalised to the signal yield extracted (with the signal strength $\mu$). The unstacked and unfilled red histogram represents the signal sample scaled by the factor indicated in the legend. The dashed histogram shows the total pre-fit background.

Table 7.8 shows the signal and background yields after the event selection, rescaled with the normalisation results of the global likelihood fit. The results are given for each category.

Figure 7.19a shows the distribution of $\log_{10}(S/B)$, constructed from the combination of the signal $S$ and the background $B$ yields of the final-discriminant bins for all the regions. The lower panel shows the statistical significance of the Higgs signal, where the red line represents the ratio between the combination of the fitted MC signal and background and the background only.

A combined fit is also performed with floating signal strengths separately for the $WH$ and $ZH$ production processes. The results of this fit are shown in Fig. 7.19b. All results are fully compatible with the SM predictions.

The total systematic uncertainty on the signal strength $\mu^{bb}_{VH}$ is directly extracted with the global likelihood fit. In order to assess the impact of the individual uncertainties introduced in Sect. 7.7, the fit is performed again with the nuisance parameter $\theta$, corresponding to a specific category, fixed to the fitted value $\hat{\theta}$ and modified upwards and downwards by its fitted uncertainty. In this fit, all the other parameters are left floating in such a way that the correlations between different uncertainties are prop-

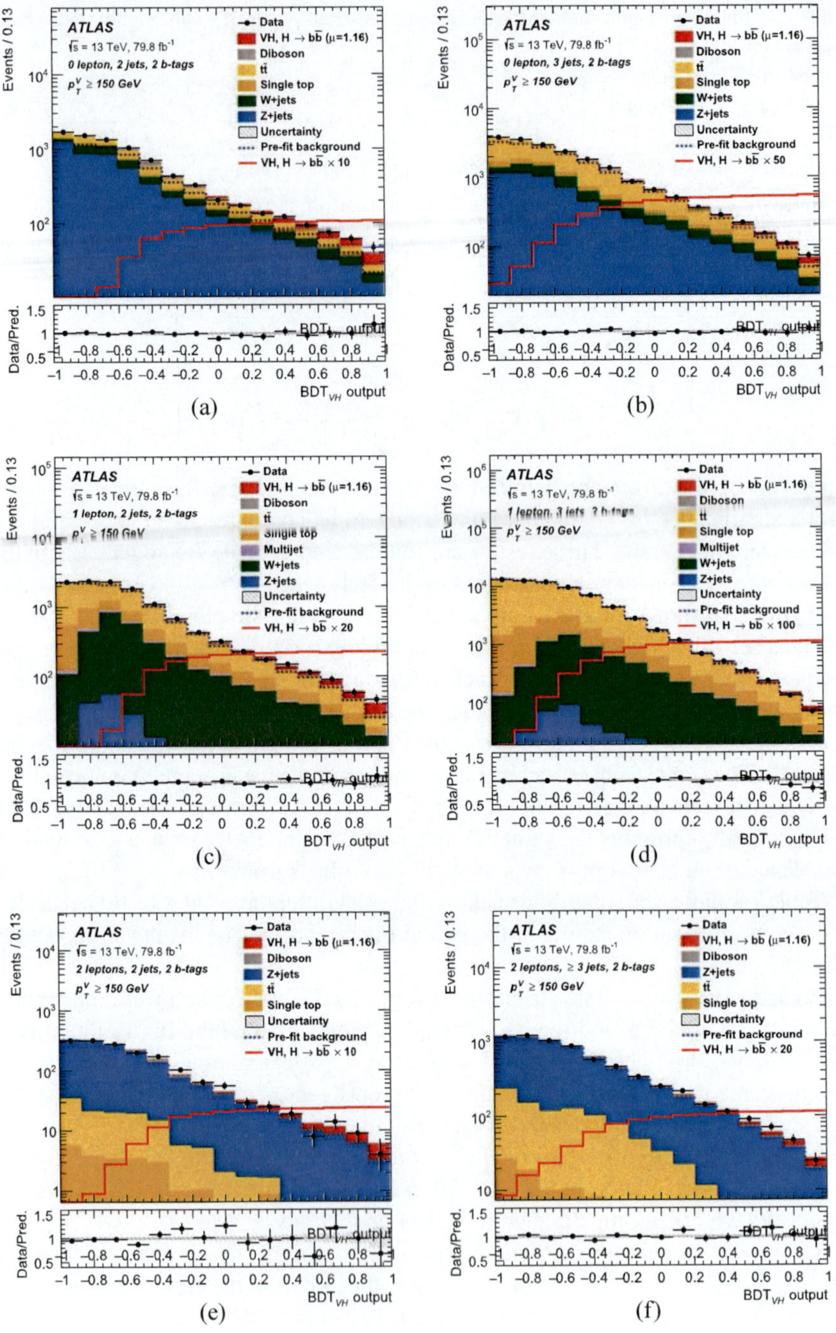


**Fig. 7.18** The BDT output post-fit distributions in the 0-lepton (top), 1-lepton (middle) and 2-lepton (bottom) channels for 2 $b$-tag events, in the 2-jet (left) and exactly 3-jet (right) categories in the high-$p_{\mathrm{T}}^{V}$ region. *Image source* [8]

**Table 7.8** Post-fit signal and background yields for each signal region category in each channel. The number of events after the selection have been rescaled with the normalisation results of the global likelihood fit. *Table source* [8]

| | 0-lepton | | 1-lepton | | 2-lepton | | | |
|---|---|---|---|---|---|---|---|---|
| | $p_T^V > 150$ GeV, 2-$b$-tag | | $p_T^V > 150$ GeV, 2-$b$-tag | | 75 GeV $< p_T^V <$ 150 GeV, 2-$b$-tag | | $p_T^V > 150$ GeV, 2-$b$-tag | |
| Process | 2-jet | 3-jet | 2-jet | 3-jet | 2-jet | ≥3-jet | 2-jet | ≥3-jet |
| $Z + ll$ | 17 ± 11 | 27 ± 18 | 2 ± 1 | 3 ± 2 | 14 ± 9 | 49 ± 32 | 4 ± 3 | 30 ± 19 |
| $Z + cl$ | 45 ± 18 | 76 ± 30 | 3 ± 1 | 7 ± 3 | 43 ± 17 | 170 ± 67 | 12 ± 5 | 88 ± 35 |
| $Z$ + HF | 4770 ± 140 | 5940 ± 300 | 180 ± 9 | 348 ± 21 | 7400 ± 120 | 14160 ± 220 | 1421 ± 34 | 5370 ± 100 |
| $W + ll$ | 20 ± 13 | 32 ± 22 | 31 ± 23 | 65 ± 48 | < 1 | < 1 | < 1 | < 1 |
| $W + cl$ | 43 ± 20 | 83 ± 38 | 139 ± 67 | 250 ± 120 | < 1 | < 1 | < 1 | < 1 |
| $W$ + HF | 1000 ± 87 | 1990 ± 200 | 2660 ± 270 | 5400 ± 670 | 2 ± 0 | 13 ± 2 | 1± 0 | 4 ± 1 |
| Single top | 368 ± 53 | 1410 ± 210 | 2080 ± 290 | 9400 ± 1400 | 188 ± 89 | 440 ± 200 | 23 ± 7 | 93 ± 26 |
| $t\bar{t}$ | 1333 ± 82 | 9150 ± 400 | 6600 ± 320 | 50200 ± 1400 | 3170 ± 100 | 8880 ± 220 | 104 ± 6 | 839 ± 40 |
| Diboson | 254 ± 49 | 318 ± 90 | 178 ± 47 | 330 ± 110 | 152 ± 32 | 355 ± 68 | 52 ± 11 | 196 ± 35 |
| Multi-jet $e$ ch. | – | – | 100 ± 100 | 41 ± 35 | – | – | – | – |
| Multi-jet $\mu$ ch. | – | – | 138 ± 92 | 260 ± 270 | – | – | – | – |
| Total bkg. | 7850 ± 90 | 19020 ± 140 | 12110 ± 120 | 66230 ± 270 | 10960 ± 100 | 24070 ± 150 | 1620 ± 30 | 6620 ± 80 |
| Signal (post-fit) | 128 ± 28 | 128 ± 29 | 131 ± 30 | 125 ± 30 | 51 ± 11 | 86 ± 22 | 28 ± 6 | 67 ± 17 |
| Data | 8003 | 19143 | 12242 | 66348 | 11014 | 24197 | 1626 | 6686 |

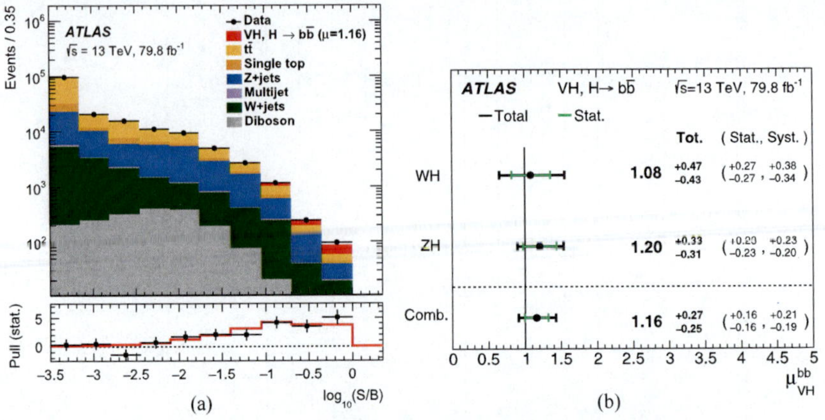


**Fig. 7.19** Figure **a** shows the event yields as a function of $\log_{10}(S/B)$ for data, background and a Higgs boson signal with $m_H = 125$ GeV. Figure **b** shows the fitted values of the Higgs boson signal strength $\mu^{bb}_{VH}$ for $m_H = 125$ GeV. The results are obtained from a simultaneous fit with the signal strength for each of the lepton channels floating independently. *Image source* [8]

erly taken into account. With this method, the variation of the signal strength with respect to the nominal result is evaluated, according to the variation of the corresponding uncertainty. The result of this procedure is shown in Fig. 7.20a, where the categories of systematics are ranked in decreasing order of their impact on $\mu^{bb}_{VH}$. The filled circles show the deviations of the fitted NPs $\hat{\theta}$ from their nominal values $\theta_0$, expressed in terms of standard deviations with respect to their nominal uncertainties $\Delta\theta$. The associated error bars show the fitted uncertainties of the nuisance parameters, relative to their nominal uncertainties.

The individual sources of uncertainties are combined (summed in quadrature) in categories whose contribution to the final measurement is provided in Fig. 7.20b. As shown in this *breakdown*, the analysis is dominated by systematic uncertainties, mostly arising from modelling of the signal, followed by the uncertainty due to the limited size of the MC samples, the modelling of the background and the $b$-tagging uncertainty.

## 7.9 Cross-Check Analyses

In order to validate the result of the multivariate analysis, two additional cross-check analyses have been performed: the analysis where the multivariate strategy is substituted by a more classic cut-based approach and the mass of the dijet system $m_{bb}$ is used as the main fit observable, and the analysis where the nominal multivariate strategy is used to extract the signal yielded by the $VZ(Z \to b\bar{b})$ diboson process.

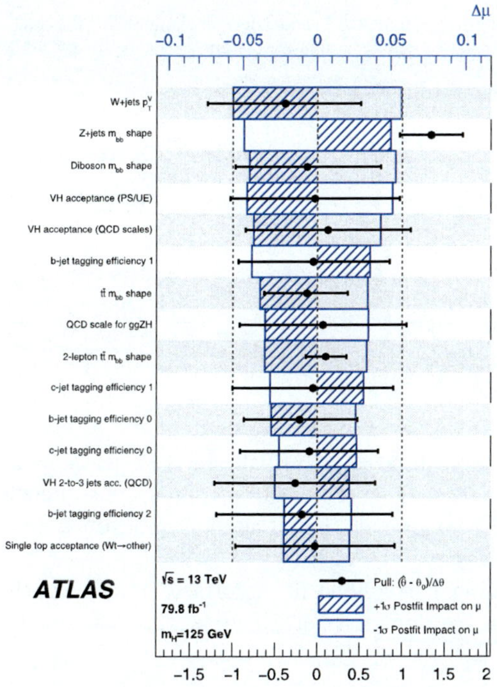


(a)

| Source of uncertainty | | $\sigma_\mu$ |
|---|---|---|
| Total | | 0.259 |
| Statistical | | 0.161 |
| Systematic | | 0.203 |
| Experimental uncertainties | | |
| Jets | | 0.035 |
| $E_T^{miss}$ | | 0.014 |
| Leptons | | 0.009 |
| | $b$-jets | 0.061 |
| $b$-tagging | $c$-jets | 0.042 |
| | light-flavour jets | 0.009 |
| | extrapolation | 0.008 |
| Pile-up | | 0.007 |
| Luminosity | | 0.023 |
| Theoretical and modelling uncertainties | | |
| Signal | | 0.094 |
| Floating normalisations | | 0.035 |
| $Z$ + jets | | 0.055 |
| $W$ + jets | | 0.060 |
| $t\bar{t}$ | | 0.050 |
| Single top quark | | 0.028 |
| Diboson | | 0.054 |
| Multi-jet | | 0.005 |
| MC statistical | | 0.070 |

(b)

**Fig. 7.20** Figure **a** show the impact of the individual systematic uncertainties on the measured signal strength $\mu_{VH}^{bb}$. Figure **b** shows the breakdown of the uncertainty in $\mu_{VH}^{bb}$ when the individual contributions are combined in categories. *Image and table source* [8]

## *7.9.1 Dijet-Mass Analysis*

The dijet-mass analysis (DMA) relies on the same object and selection criteria described in Sect. 7.3 for the main MVA analysis, with the inclusion of additional cuts, listed in Table 7.9, which help to reduce the background contamination. A cut on the transverse mass $m_T^W$ in 1-lepton and a cut on the $E_T^{miss}$ significance $E_T^{miss}/\sqrt{S_T}$ in the 2-lepton channel allow us to reduce the $t\bar{t}$ background. The high-$p_T^V$ region is further split into two regions $150 < p_T^V < 200$ GeV and $p_T^V > 200$ GeV. A cut on the angular separation between the two $b$-tagged jets $\Delta R(\vec{b}_1, \vec{b}_2)$ is applied with different values according to the $p_T^V$ region. In fact, for a Higgs boson splitting into two $b$-jets with approximately same momentum, its mass is related to the object separation as

$$\Delta R(\vec{b}_1, \vec{b}_2) \sim \frac{m_H}{p_T^H} \tag{7.11}$$

**Table 7.9** Summary of the event selection criteria in the 0-, 1- and 2-lepton channels for the dijet-mass analysis, applied in addition to those described in Sect. 7.3 for the multivariate analysis. *Table source* [8]

| Channel | | | |
|---|---|---|---|
| Selection | 0-lepton | 1-lepton | 2-lepton |
| $m_{\mathrm{T}}^{W}$ | – | < 120 GeV | – |
| $E_{\mathrm{T}}^{\mathrm{miss}}/\sqrt{S_{\mathrm{T}}}$ | – | – | $< 3.5\sqrt{\mathrm{GeV}}$ |
| $p_{\mathrm{T}}^{V}$ regions | | | |
| $p_{\mathrm{T}}^{V}$ | 75 – 150 GeV (2-lepton only) | 150 – 200 GeV | > 200 GeV |
| $\Delta R(\vec{b}_1, \vec{b}_2)$ | <3.0 | <1.8 | <1.2 |

where, the larger is the transverse momentum $p_{\mathrm{T}}^{H}$, the smaller is the $b$-jets separation [35]. Since in the 1-lepton channel the low $m_{bb}$ dijet mass provides sufficient information to constrain the $W$+ HF background normalisation, the construction of a dedicated control region is not necessary.

The signal is extracted with a global likelihood fit performed simultaneously across the three lepton channels on the dijet-mass distribution. The value of the signal strength is found to be:

$$\mu_{VH}^{bb} = 1.06 \pm 0.20(\mathrm{stat.})^{+0.30}_{-0.26}(\mathrm{syst.}) = 1.06^{+0.36}_{-0.33}.$$

This result, summarised in Fig. 7.21a, is in agreement with the main MVA analysis and corresponds to an observed (expected) significance of 3.6 (3.5) standard deviations. With a 27% smaller significance, the dijet analysis is less sensitive than the nominal MVA analysis. In Fig. 7.21b is shown the $m_{bb}$ distribution obtained by subtracting the background contribution, except for the $WZ$ and $ZZ$ diboson processes, from the signal of each analysis bin and weighting it by the ratio of fitted signal and background in each bin.

### 7.9.2 *Diboson Analysis*

The nominal MVA strategy is validated with a diboson analysis by measuring the signal strength of the $VZ$ processes, instead of $VH$. A new $\mathrm{BDT}_{VZ}$ is trained to separate the $VZ(Z \to b\bar{b})$ signal from all the other background processes, including the SM $VH(H \to b\bar{b})$. A global likelihood fit returns a value of signal strength

$$\mu_{VZ}^{bb} = 1.20 \pm 0.08(\mathrm{stat.})^{+0.19}_{-0.16}(\mathrm{syst.}) = 1.20^{+0.20}_{-0.18},$$

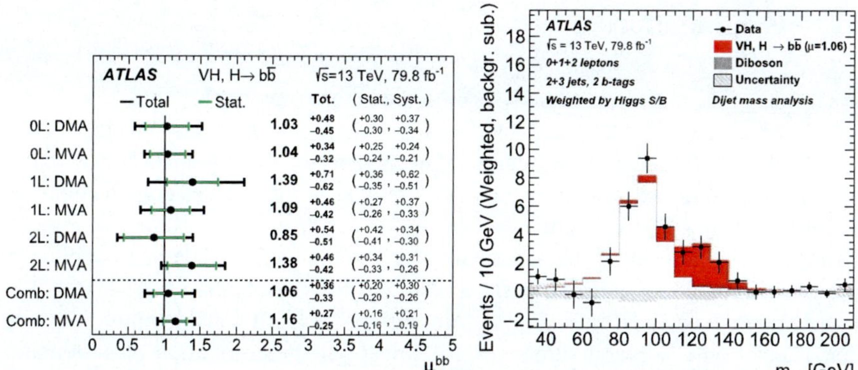


(a) (b)

**Fig. 7.21** Fitted values of the Higgs boson signal strength $\mu_{VH}^{bb}$ for the 0-, 1- and 2-lepton channels and their combination. The results are shown both for the nominal MVA and for the di-jet mass analysis (DMA) (**a**). $m_{bb}$ distribution obtained from the DMA analysis (**b**). *Image source* [8]

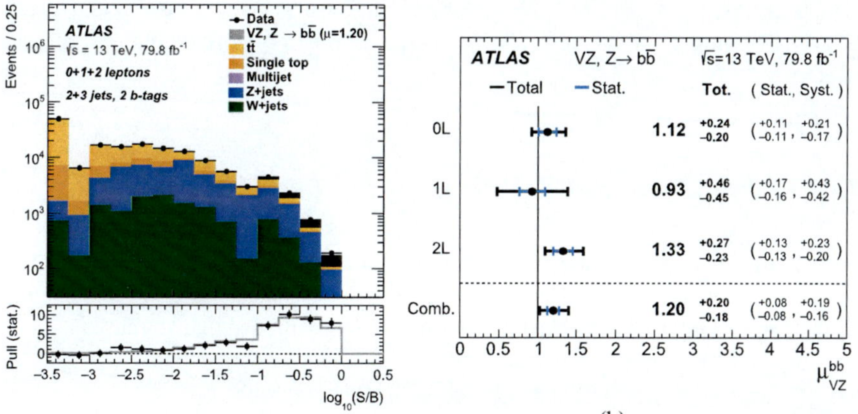


(a) (b)

**Fig. 7.22** Figure **a** shows the event yields as a function of $\log_{10}(S/B)$ for data, background and diboson ($VZ$) signal. Figure **b** shows the fitted values of the $VZ$ signal strength for the 0-lepton, 1-lepton and 2-lepton channels. The results are obtained from a simultaneous fit with the signal strength for each of the lepton channels floating independently. *Image source* [8]

which corresponds to a signal significance much larger than $5\sigma$. Similarly to the $VH(H \to b\bar{b})$ MVA analysis, this result can be visualised with the $\log_{10}(S/B)$ distribution, as shown in Fig. 7.22a. Figure 7.22b summarises the results of simultaneous fit where the individual signal strengths for each of the lepton channels are left floating independently.

## 7.10 Combinations

The results of the multivariare analysis presented in the previous sections, which was performed with 13 TeV data collected during Run-2, have been combined with the Run-1 dataset [36], recorded at a centre of mass energy of 7 TeV and 8 TeV, with the aim to improve the sensitivity of the measurement. Furthermore, combinations of analyses that are sensitive to the $H \to b\bar{b}$ decay or $VH$ production mode have been performed and have led to the observation of such processes. The main challenge in these combinations is to properly understand the correlation of the systematic uncertainties, especially those arising from different experimental configurations of Run-1 and Run-2. Several studies on the impact of the correlation of systematic uncertainties between the $VH(H \to b\bar{b})$ analyses performed with different centre-of-mass energies were conducted in the 2017 analysis [23]. Most of the NPs have negligible correlations and only the JES for $b$-jets and theory uncertainties in the signal process are found to be correlated. Potential correlations in the background modelling, due to different cross section and acceptance, different MC generators, as well as different reconstruction, selection and analysis procedure, have been studied as well. The impact of such correlations is found to be no more than a few percent effect.

### 7.10.1 Run-1 and Run-2 $VH(H \to b\bar{b})$ Combination

The combination of the $VH(H \to b\bar{b})$ analyses conducted in Run-1 and Run-2 produces a p-value of $5.5 \cdot 10^{-7}$, corresponding to an observed (expected) signal significance of 4.9 (5.1) standard deviations. The measured signal strength is:

$$\mu_{VH}^{bb} = 0.98 \pm 0.14(\text{stat.})^{+0.17}_{-0.16}(\text{syst.}) = 0.98^{+0.22}_{-0.21}.$$

The results of multiple parameter-of-interest fit, obtained with independent floating signal strength for the $WH$ and $ZH$, is shown in Fig. 7.23.

### 7.10.2 $H \to b\bar{b}$ Observation

The combination is based on the analyses of the Higgs boson production modes conducted in Run-1 and Run-2. The following analyses have been included in the combination:

**$VH(H \to b\bar{b})$** The Run-1 analysis performed with 4.7 fb$^{-1}$ at a centre-of-mass energy of 7 TeV and with 20.3 fb$^{-1}$ at 8 TeV [36]. The Run-2 analysis used in the combination is the one presented in this chapter, performed with 78.9 fb$^{-1}$ at a centre-of-mass energy of 13 TeV [8].

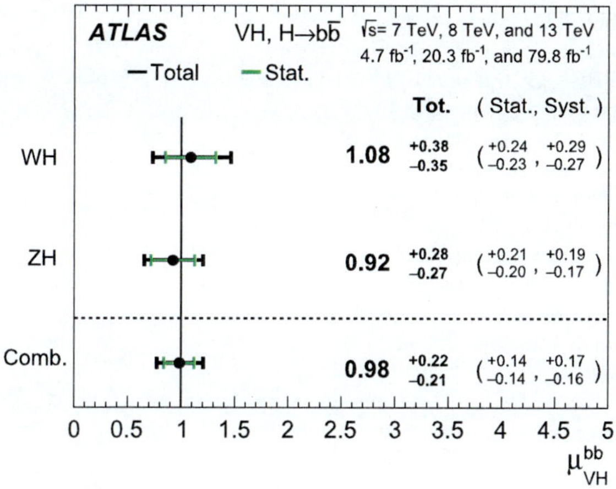


**Fig. 7.23** Higgs boson signal strength $\mu_{VH}^{bb}$ for the $WH$ and $ZH$ processes in the combination of Run-1 and Run-2, using the 7 TeV, 8 TeV and 13 TeV data. The displayed values are the result of a global likelihood fit where the individual signal strengths of $WH$ and $ZH$ are left floating. *Image source* [8]

**VBF** The VBF production mode with the Higgs boson decaying into a pair of bottom quarks is targeted by the analyses of Run-1, conducted with 20.2 fb$^{-1}$ at 8 TeV [37], and Run-2, conducted with 30.6 fb$^{-1}$ at 13 TeV [38]. Since a non-negligible fraction of gluon-gluon fusion events are present, these analyses are referred to as VBF+ggF.

***ttH*** The analyses of the Higgs production in association with a pair of top quarks have been used in the combination. In Run-1 the analysis was conducted with a dataset of 20.3 fb$^{-1}$ at 8 TeV [39], while in Run-2 it was performed with 36.1 fb$^{-1}$ at 13 TeV [40].

The combined fit performed as a search of the $H \to b\bar{b}$ decay results in a signal strength

$$\mu_{VH}^{bb} = 1.01 \pm 0.12(\text{stat.})^{+0.16}_{-0.15}(\text{syst.}) = 1.01 \pm 0.20.$$

The observed significance for the $H \to b\bar{b}$ decay is 5.4 standard deviations, while the expectation is 5.5 standard deviations. The combination of Run-1 and Run-2 different production mode analyses provides the necessary sensitivity for the observation of the $H \to b\bar{b}$ decay. The results of the global likelihood fit performed with floating signal strength parameters separately for VBF+ggF, $ttH$ and $VH$, are shown in Fig. 7.24.

### *7.10.3 VH Observation*

Another combination is carried out with Run-2 measurements, in order to improve the sensitivity of the specific $VH$ production channel. The following decay channels have been exploited in the combination:

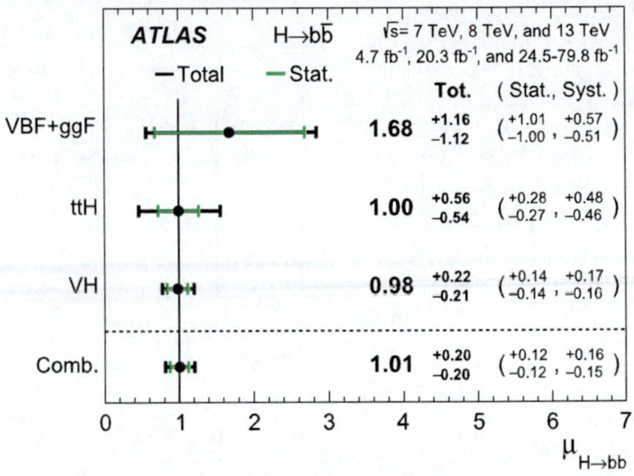


**Fig. 7.24** Higgs boson signal strength $\mu_{VH}^{bb}$ for the VBF+ggF, $ttH$ and $VH$ production modes and Higgs boson decaying into a pair of bottom quarks. The displayed values are the result of a global likelihood fit where the individual signal strengths of the different production modes are left floating. *Image source* [8]

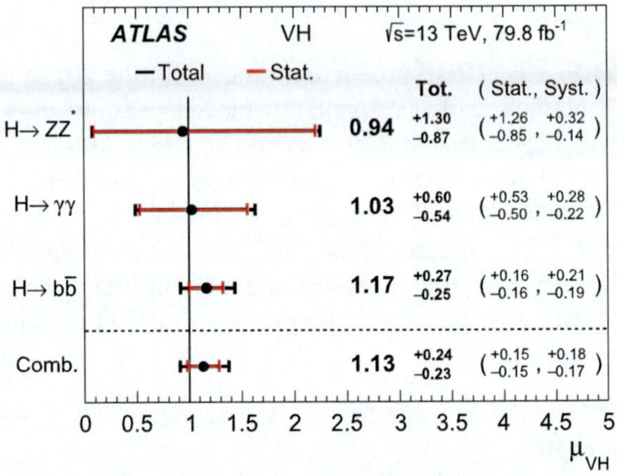


**Fig. 7.25** Higgs boson signal strength $\mu_{VH}^{bb}$ for the $H \rightarrow b\bar{b}$, $H \rightarrow ZZ^*$ and $H \rightarrow \gamma\gamma$ decay modes and $VH$ production. The displayed values are the result of a global likelihood fit where the individual signal strengths of the different decay modes are left floating. *Image source* [8]

***$H \rightarrow bb$*** The process analysed in this thesis work, performed during Run-2 with a dataset corresponding to 79.8 fb$^{-1}$.

***$H \rightarrow \gamma\gamma$*** The corresponding measurement has been performed with a Run-2 dataset corresponding to 79.8 fb$^{-1}$, whose strategy is documented in [41].

***$H \rightarrow ZZ^* \rightarrow 4\ell$*** The measurement of the Higgs boson in the $VH$ production mode decaying into leptons, either electrons or muons, have been performed with 79.8 fb$^{-1}$ [42].

The combined fit performed as a search of the $VH$ production mode results in a signal strength

$$\mu_{VH}^{bb} = 1.13 \pm 0.15(\text{stat.})^{+0.18}_{-0.17}(\text{syst.}) = 1.13^{+0.24}_{-0.23}.$$

The observed significance for the $VH$ production channel is 5.3 standard deviations, while the expectation is 4.8 standard deviations, thus providing the necessary sensitivity to claim the observation of the $VH$ production mode. The results of the global likelihood fit performed with floating signal strength parameters separately for $H \to b\bar{b}$, $H \to ZZ^*$ and $H \to \gamma\gamma$, are shown in Fig. 7.25.

## References

1. Aad G et al (2012) Observation of a new particle in the search for the Standard Model Higgs boson with the ATLAS detector at the LHC. Phys Lett B716:1–29. https://doi.org/10.1016/j.physletb.2012.08.020. arXiv: 1207.7214 [hep-ex]
2. Chatrchyan S et al (2012) Observation of a New Boson at a Mass of 125 GeV with the CMS Experiment at the LHC. Phys Lett B716:30–61. https://doi.org/10.1016/j.physletb.2012.08.021. arXiv: 1207.7235 [hep-ex]
3. Measurements of the Higgs boson production, fiducial and differential cross sections in the $4l$ decay channel at $\sqrt{s} = 13\ TeV$ with the ATLAS detector. Technical report ATLAS-CONF-2018-018. Geneva: CERN (2018). https://cds.cern.ch/record/2621479
4. Aad G et al (2015) Observation and measurement of Higgs boson decays to WW* with the ATLAS detector. Phys Rev D92.1:012006. https://doi.org/10.1103/PhysRev.D.92.012006. arXiv: 1412.2641 [hep-ex]
5. Aad G et al (2014) Measurement of the Higgs boson mass from the $H \to \gamma\gamma$ and $H \to ZZ^* \to 4l$ channels in pp collisions at center-of-mass energies of 7 and 8 TeV with the ATLAS detector. Phys Rev D 90(5):052004. https://doi.org/10.1103/PhysRevD.90.052004. https://link.aps.org/doi/10.1103/PhysRevD.90.052004
6. Aad G et al (2016) Measurements of the Higgs boson production and decay rates and constraints on its couplings from a combined ATLAS and CMS analysis of the LHC pp collision data at $\sqrt{s} = 7$ and 8 TeV. JHEP 08:045. https://doi.org/10.1007/JHEP08(2016)045. arXiv: 1606.02266 [hep-ex]
7. Aaboud M et al (2018) Observation of Higgs boson production in association with a top quark pair at the LHC with the ATLAS detector. Phys Lett B784:173-191. https://doi.org/10.1016/j.physletb.2018.07.035. arXiv: 1806.00425 [hep-ex]
8. ATLAS Collaboration (2018) Observation of $H \to b\bar{b}$ decays and $VH$ production with the ATLAS detector. Phys Lett B786:59-86. https://doi.org/10.1016/j.physletb.2018.09.013. arXiv: 1808.08238 [hep-ex]
9. de Florian D et al (2016) Handbook of LHC Higgs cross sections: 4. Deciphering the nature of the Higgs sector. In: (2016). https://doi.org/10.23731/CYRM-2017-002. arXiv: 1610.07922 [hep-ph]
10. Stirling WJ, Parton luminosity and cross section plots. http://www.hep.ph.ic.ac.uk/~wstirlin/plots/plots.html
11. Aaltonen T et al (2012) Evidence for a particle produced in association with weak bosons and decaying to a bottom-antibottom quark pair in Higgs boson searches at the Tevatron. Phys Rev Lett 109:071804. https://doi.org/10.1103/PhysRevLett.109.071804. arXiv:1207.6436 [hep-ex]
12. Aaboud M et al (2017) Evidence for the $H \to b\bar{b}$ decay with the ATLAS detector. JHEP 12:024. https://doi.org/10.1007/JHEP12(2017)024. arXiv: 1708.03299 [hep-ex]
13. Sirunyan AM et al (2018) Evidence for the Higgs boson decay to a bottom quark-antiquark pair. Phys Lett B780:501–532. https://doi.org/10.1016/j.physletb.2018.02.050. arXiv: 1709.07497 [hep-ex]
14. Alioli S et al (2010) A general framework for implementing NLO calculations in shower Monte Carlo programs: the POWHEG BOX. JHEP 06:043. https://doi.org/10.1007/JHEP06(2010)043. arXiv: 1002.2581 [hep-ph]
15. Sjostrand T, Mrenna S, Skands PZ (2008) A Brief Introduction to PYTHIA 8.1. Comput Phys Commun 178:852–867. https://doi.org/10.1016/j.cpc.2008.01.036. arXiv: 0710.3820 [hep-ph]

16. Gleisberg T et al (2009) Event generation with SHERPA 1.1. JHEP 02:007. https://doi.org/10.1088/1126-6708/2009/02/007. arXiv: 0811.4622 [hep-ph]
17. Agostinelli S et al (2003) Geant4-a simulation toolkit. Nuclear Instrum Methods Phys Res Sect A: Accel Spectr Detect Assoc Equip 506(3):250–303. ISSN: 0168-9002. https://doi.org/10.1016/S0168-9002(03)01368-8. http://www.sciencedirect.com/science/article/pii/S0168900203013688
18. Calafiura P et al (2018) The Athena control framework in production, new developments and lessons learned. In: (2005). https://doi.org/10.5170/CERN-2005-002.456. https://cds.cern.ch/record/865624
19. Ahmadov F et al (2018) Search for the Standard Model Higgs boson produced in association with a vector boson and decaying to a pair of b-quarks. Technical report. ATL-COM-PHYS-2018- 512. Geneva: CERN. https://cds.cern.ch/record/2317111
20. Andersen JR et al (2013) Handbook of LHC Higgs cross sections: 3. Higgs properties. In: (2013). Heinemeyer S et al (eds). https://doi.org/10.5170/CERN-2013-004. arXiv: 1307.1347 [hep-ph]
21. Buzatu A, Jiggins S (2018) Object definitions and selections for standard model $VH \to ll/l\nu/\nu\nu + b\bar{b}$ Analysis. Technical report. ATL-COM-PHYS-2018-517. Geneva: CERN, May 2018. https://cds.cern.ch/record/2317182
22. Aaboud M et al (2018) Measurements of b-jet tagging efficiency with the ATLAS detector using $t\bar{t}$ events at $\sqrt{s} = 13$ TeV. JHEP 08:089. https://doi.org/10.1007/JHEP08(2018)089. arXiv: 1805.01845 [hep-ex]
23. Aaboud M et al (2018) Evidence for the $H \to b\bar{b}$ decay with the ATLAS detector. JHEP 12:024. https://doi.org/10.1007/JHEP12(2017)024. arXiv: 1708.03299 [hep-ex]
24. Zyla PA, Others (Particle Data Group) (2020) Review of particle physics. PTEP 2020(8):083C01. https://doi.org/10.1093/ptep/ptaa104
25. Proissl M (2014) Dijet invariant mass studies in the Higgs boson $H \to b\bar{b}$ resonance search in association with a $W/Z$ boson using the ATLAS detector. Presented 2014. https://cds.cern.ch/record/1997808
26. Bukin AD (2007) Fitting function for asymmetric peaks. arXiv e-prints, arXiv:0711.4449 [physics.data-an]
27. Hocker A et al (2007) TMVA - toolkit for multivariate data analysis with ROOT: users guide. Technical report Geneva: CERN. https://cds.cern.ch/record/1019880
28. James G et al (2014) An introduction to statistical learning: with applications in R. Springer Publishing Company, Incorporated. ISBN: 1461471370
29. Friedman J (2018) Greedy function approximation: a gradient boosting machine. Ann Stat 29. https://doi.org/10.1214/aos/1013203451
30. Brownlee J, A gentle introduction to k-fold cross-validation. https://machinelearningmastery.com/k-fold-cross-validation/
31. Andreas Hocker et al (2007) TMVA - toolkit for multivariate data analysis. In: (2007). arXiv: physics/0703039 [physics.data-an]
32. Cowan G et al (2011) Asymptotic formulae for likelihood-based tests of new physics. Eur Phys J C71 (2011). [Erratum: Eur Phys J C73,2501(2013)] 1554. https://doi.org/10.1140/epjc/s10052-011-1554-0, https://doi.org/10.1140/epjc/s10052-013-2501-z. arXiv: 1007.1727 [physics.data-an]
33. Barlow RJ, Christine B (1993) Fitting using finite Monte Carlo samples. Comput Phys Commun 77:219–228. https://doi.org/10.1016/0010-4655(93)90005-W
34. van der Meer S (1968) Calibration of the effective beam height in the ISR. Technical report. CERN-ISR-PO-68-31. ISR-PO-68-31. Geneva: CERN. https://cds.cern.ch/record/296752
35. Salam GP (2018) Towards jetography. Eur Phys J C67:637–686. https://doi.org/10.1140/epjc/s10052-010-1314-6. arXiv: 0906.1833 [hep-ph]
36. Search for the bb decay of the Standard Model Higgs boson in associated W/ZH production with the ATLAS detector. Technical report. ATLAS-CONF-2013-079. Geneva: CERN (2013). https://cds.cern.ch/record/1563235

37. Aaboud M et al (2016) Search for the Standard Model Higgs boson produced by vector-boson fusion and decaying to bottom quarks in $\sqrt{s} = 8$ TeV pp collisions with the ATLAS detector. JHEP 11:112. https://doi.org/10.1007/JHEP11(2016)112. arXiv:1606.02181 [hep-ex]
38. Aaboud M et al (2018) Search for Higgs bosons produced via vector-boson fusion and decaying into bottom quark pairs in $\sqrt{s} = 13$ TeV pp collisions with the ATLAS detector. Phys Rev D98(5):052003. https://doi.org/10.1103/PhysRevD.98.052003. arXiv: 1807.08639 [hep-ex]
39. Georges Aad et al (2016) Search for the Standard Model Higgs boson decaying into $b\bar{b}$ produced in association with top quarks decaying hadronically in pp collisions at $\sqrt{s} = 8$ TeV with the ATLAS detector. JHEP 05:160. https://doi.org/10.1007/JHEP05(2016)160. arXiv: 1604.03812 [hep-ex]
40. Aaboud M et al (2018) Search for the standard model Higgs boson produced in association with top quarks and decaying into a $b\bar{b}$ pair in pp collisions at $\sqrt{s} = 13$ TeV with the ATLAS detector. Phys Rev D97(7):072016. https://doi.org/10.1103/PhysRevD.97.072016. arXiv: 1712.08895 [hep-ex]
41. Aaboud M et al (2018) Measurements of Higgs boson properties in the diphoton decay channel with 36 $\text{fb}^{-1}$ of pp collision data at $\sqrt{s} = 13$ TeV with the ATLAS detector. Phys Rev D98:052005. https://doi.org/10.1103/PhysRevD.98.052005. arXiv: 1802.04146 [hep-ex]
42. Aaboud M et al (2018) Measurement of the Higgs boson coupling properties in the $H \to ZZ^* \to 4l$ decay channel at $\sqrt{s} = 13$ TeV with the ATLAS detector. JHEP 03:095. https://doi.org/10.1007/JHEP03(2018)095. arXiv: 1712.02304 [hep-ex]

# Chapter 8
# $VH, H \to b\bar{b}$ Cross Sections and Effective Field Theory

## 8.1 Cross Section Measurements

LHC Run-1 and Run-2 were mainly devoted to probing the SM predictions on the Higgs processes with signal strength measurements that exploit at best the experimental sensitivity, as in the $VH(H \to b\bar{b})$ analysis discussed in Chap. 7. On the other hand, as the experimental measurements have become more precise, phenomenological searches for deviations from the SM predictions have been enabled. This kind of measurement requires the definition of a common framework where these deviations can be computed precisely and in a fashion that is as independent of the theory as possible [1].

The signal strength for a generic production mode $i$ and decay $f$ takes the form

$$\mu_i^f = \frac{(\sigma_i \cdot \mathrm{BR}^f)^{\mathrm{exp}}}{(\sigma_i \cdot \mathrm{BR}^f)^{\mathrm{theo}}}. \tag{8.1}$$

This quantity is strongly dependent on the underlying theoretical assumptions used to predict $(\sigma_i \cdot \mathrm{BR}^f)^{\mathrm{theo}}$ and therefore is not the recommended approach to present results of precise measurements of the Higgs boson. An alternative choice is provided by the *total cross section*, which is defined by quantities that are directly measurable:

$$(\sigma_i)_{\mathrm{exp}} = \frac{N_{\mathrm{ev},if}}{\mathrm{BR}^f \mathcal{A}_{if} \epsilon_{if} \mathcal{L}}. \tag{8.2}$$

Apart from the luminosity $\mathcal{L}$ and the number of observed events $N_{\mathrm{ev},if}$, the measured quantities, such as the acceptance $\mathcal{A}$, the experimental efficiency $\epsilon$ and the branching ratio BR, still depend on theoretical assumptions. A requirement for the experimental quantities is that they should be directly comparable with the theory. This is achievable by implementing the event selection at generator level on simulated objects that have not interacted yet with the simulated detector. These selections identify *fiducial*



C. Tosciri, *Higgs Boson Decays into a Pair of Bottom Quarks*, Springer Theses,
https://doi.org/10.1007/978-3-030-87938-9_8

*volumes* on the truth level phase space and are defined to resemble the experimental selection and the detector acceptance cuts [2]. In general, not all the events falling in a fiducial volume are properly reconstructed at detector level and some corrections are required, which introduce a theoretical dependence. The procedure used to correct experimental effects, such as acceptance and resolution effects, is called *unfolding* into fiducial regions. The larger is the fraction of the reconstructed events from the fiducial volume, the smaller is the theoretical contribution to the correction. Therefore, with a careful definition of fiducial volumes, the dependence to the model can be minimised. All these considerations bring to the definition of *fiducial cross sections with fixed final state* which do not require extrapolations to the full phase space [1]:

$$(\sigma_i \times \text{BR}^f \times \mathcal{A}_{if})_{\text{exp}} = \frac{N_{\text{ev},if}}{\epsilon_{if}\mathcal{L}}. \tag{8.3}$$

Typically, the best choice to achieve a measurement least dependent on the assumed theoretical model is represented by *differential fiducial cross sections*, which are fiducial cross sections with fixed final state, measured as a function of event kinematic observables.

The fiducial and measured cross sections, respectively $\sigma_i^{\text{fid}}$ and $\sigma_j$, are related through an *unfolding* or *migration matrix* $\mathcal{A}_{ij}$ as

$$\sigma_i^{\text{fid}} = \sum_j \mathcal{A}_{ij}\sigma_j, \tag{8.4}$$

where $i$ and $j$ define the bin where a particular differential cross section is measured.

Despite being the recommended approach for comparing a measurement to a theoretical prediction, a differential fiducial measurement is not always feasible as it requires high statistics, as well as very clean decay channels and production modes. Furthermore, the extraction of differential fiducial cross sections is usually enabled by event selections based on kinematic cuts, while many Higgs analyses rely on multivariate techniques to enhance their sensitivity, like the $VH(H \to b\bar{b})$ analysis.

## 8.2 Simplified Template Cross Section Framework

A new common framework has been developed by the LHC community to combine the advantages of the signal strength measurements and the fiducial and differential measurements [3, 4]. The simplified template cross section (STXS) framework allows the measurement of cross sections defined in exclusive regions of phase space. This reduces significantly the model dependence with respect to a signal strength measurement and maximises the experimental sensitivity. The definition of the simplified fiducial volumes, in fact, allows the use of advanced multivariate techniques. The cross section $\sigma_{\text{meas}}$ measured in a specific Higgs production and decay mode is related to the STXSs as

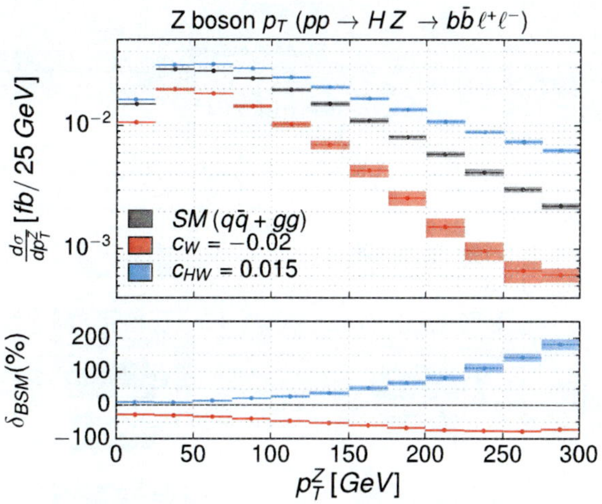


**Fig. 8.1** Transverse momentum of the $Z$ boson produced in association with a Higgs boson. The SM differential distribution (grey) is compared to the predicted distribution obtained assuming that anomalous coefficients could modify the Higgs couplings with the vector boson (red and blue). Image source: [5]

$$\begin{aligned}
\sigma_{\text{meas}} &= A_{\alpha_1}\sigma_{\alpha_1} \\
\sigma_{\text{meas}} &= A_{\beta_1}\sigma_{\beta_1} + A_{\beta_2}\sigma_{\beta_2} \\
&\ldots \\
\sigma_{\text{meas}} &= A_{\gamma_1}\sigma_{\gamma_1} + A_{\gamma_2}\sigma_{\gamma_2} + \cdots + A_{\gamma_n}\sigma_{\gamma_n},
\end{aligned} \tag{8.5}$$

where the $\alpha_k$ ($\beta_k$, $\gamma_k$) indicate the splitting in $k$ fiducial regions, the $\sigma_{\alpha_k(\beta_k,\gamma_k)}$ are the STXSs and the $A_{\alpha_k(\beta_k,\gamma_k)}$ are factors that account for the efficiency and acceptance in each STXS. The STXS framework is defined for progressive *stages* and each line of Equation (8.5) corresponds to a stage which is characterised by an increasing splitting of the fiducial space, resulting in multiple *bins*. The definition of the STXS bins follows the criteria of minimisation of the theoretical dependence and maximisation of the experimental sensitivity within each bin. Potential effects from BSM physics are expected to be isolated in specific regions of the phase space, as shown in Fig. 8.1.

The framework is constructed in such a way that it takes advantage, in terms of precision of the measurement, from combinations, and at the same time aims at maximising the granularity of the measurement. The bins must be mutually exclusive in order to avoid statistical correlations between adjacent bins. The splitting is determined by kinematic observables that are directly measured by the experiments and reduces the extrapolation into larger regions of phase space.

In the first stage of the STXS framework, called *stage-0*, the cross sections are measured in the Higgs boson rapidity region $|y_H| < 2.5$, within single inclusive bins, each corresponding to one of the main Higgs production modes: $gg$F, VBF, $VH$ and $t\bar{t}H$. At *stage-1*, a number of subregions are obtained by further splitting those in stage-0. The splitting is determined by kinematic properties such as the number of selected jets, the transverse momentum of the Higgs boson, or the transverse

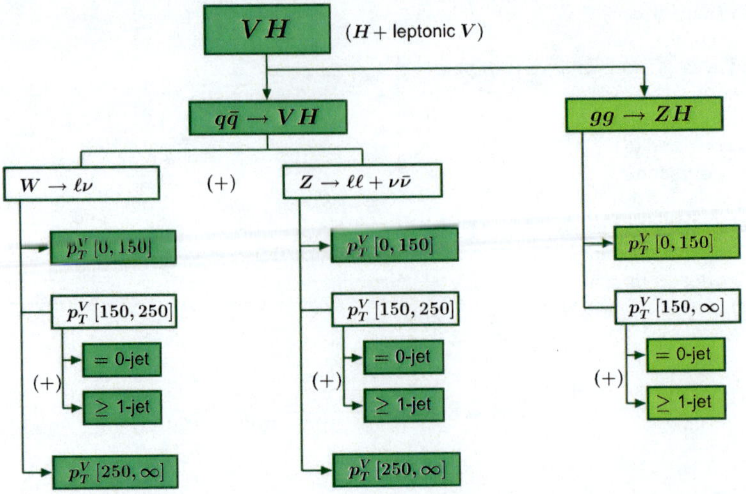


Fig. 8.2 STXS stage-1 binning for associated production with vector bosons. Image source: [3]

momentum of the vector boson $V$, in case of $VH$ production. An example of stage-1 splitting for the $VH$ productions is shown in Fig. 8.2. The first split of the $VH$ production, based on the initial state production, $qq$ or $gg$, is defined according to the MC samples used in the analysis. Moreover, $qq$ is split according to the vector boson, $W \to \ell\nu$ and $Z \to \ell\ell/\nu\nu$, and then into bins of $p_{\mathrm{T}}^{V}$.

## 8.3 $VH(H \to b\bar{b})$ STXS Measurements

The $VH(H \to b\bar{b})$ analysis presented in Chap. 7 has been extended to provide a measurement of the cross sections as a function of the gauge boson transverse momentum, using the STXS framework [6]. The STXS bins are defined by fiducial selections according to kinematic properties of the Higgs production, following a modified version of stage-1 shown in Fig. 8.2, called *reduced stage-1*. The $qq$ and $gg$ initial states for the $ZH$ production can not be distinguished by the current analysis, due to insufficient statistics in data. In order to reflect the analysis acceptance at the reconstruction level, the $p_T^Z < 150$ GeV stage-1 regions are split into two subregions: $p_T^Z < 75$ GeV and $75 < p_T^Z < 150$ GeV. Two sets of reduced stage-1 splittings, shown in Fig. 8.3, are employed to measure three or five cross sections, in the schemes respectively defined as '5-POI' and '3-POI', according to the number of parameters of interest (POI).

The 5-POI scheme defines three bins for $ZH$ production ($75 < p_{\mathrm{T}}^{Z} < 150$ GeV, $150 < p_{\mathrm{T}}^{Z} < 250$ GeV and $p_{\mathrm{T}}^{Z} > 250$ GeV) and two for $WH$ production ($150 <$

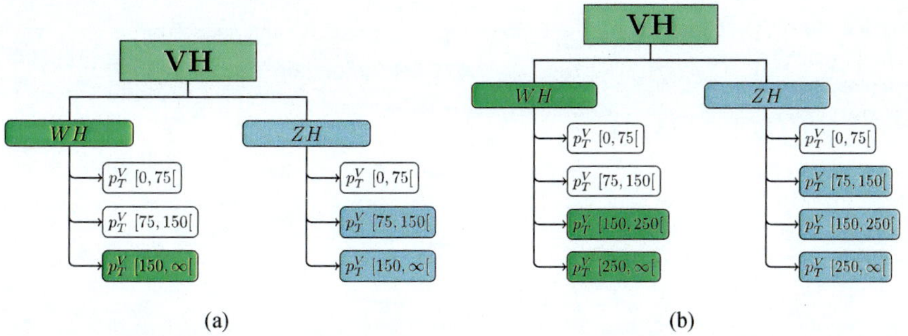


**Fig. 8.3** Figure 8.3a shows the 3-POI scheme used for the $VH(H \to b\bar{b})$ cross section measurements. Similarly, Fig. 8.3b shows a finer bin splitting used to perform 5-POI cross section measurements. Image source: [6]

$p_{\mathrm{T}}^{W} < 250$ GeV and $p_{\mathrm{T}}^{W} > 250$ GeV). In the 3-POI case, two cross sections for $ZH$ ($75 < p_{\mathrm{T}}^{Z} < 150$ GeV and $p_{\mathrm{T}}^{Z} > 150$ GeV) and one for $WH$ ($p_{\mathrm{T}}^{W} > 150$ GeV), are measured. The resolution on $p_{\mathrm{T}}^{V}$ varies between a few percent in case of 2-lepton channel, and ~10% in the 0-lepton channel.

Apart from dedicated assessment of theoretical systematics, the strategy of the analysis, including the event selection and categorisation, as well as the use of a BDT that maximises the analysis sensitivity, remains unchanged with respect to the one presented in Chap. 7. The event categorisation is not explicitly designed to distinguish between events with $p_{\mathrm{T}}^{V}$ below or above 250 GeV. The sensitivity to the two regions targeted by the STXS approach is provided by the different shape assumed by the BDT discriminant used in the final fit to data. The BDT shapes are shown in Fig. 8.4 separately for the two $p_{\mathrm{T}}^{V}$ regions in the 1-lepton, 2jet category.

A maximum likelihood fit is performed simultaneously on the STXS regions. The fit model is very similar to the one used for the $VH(H \to b\bar{b})$ analysis and discussed in Sects. 7.6 and 7.7, with a few differences in the treatment of the signal. In the cross section measurements, multiple shapes and yields, according to the number of regions under study, substitute the single shape or yield used in each category for the signal strength measurement, and multiple parameters of interest now replace the single parameter of interest of the fit. Additionally, theoretical uncertainties associated with the signal strength do not enter anymore the fit model. The systematic uncertainties are re-evaluated for the signal samples in each STXS region. The $VH$ cross sections times the $H \to b\bar{b}$ branching ratio are measured on each of the three or five STXS regions and summarised in Table 8.1.

The templates for the 3-POI scheme in the regions with $p_{\mathrm{T}}^{V} > 150$ GeV are obtained by merging the templates in the regions with $150 < p_{\mathrm{T}}^{V} < 250$ GeV and $p_{\mathrm{T}}^{V} > 250$ GeV of the 5-POI scheme, and assuming that the ratio of yields in those regions has the value predicted by the SM. The relative uncertainties of the measurements are found to be between 29 and 56% in the 3-POI scheme and between 56 and 125% in 5-POI case. This is mainly driven by statistical uncertainties, which become

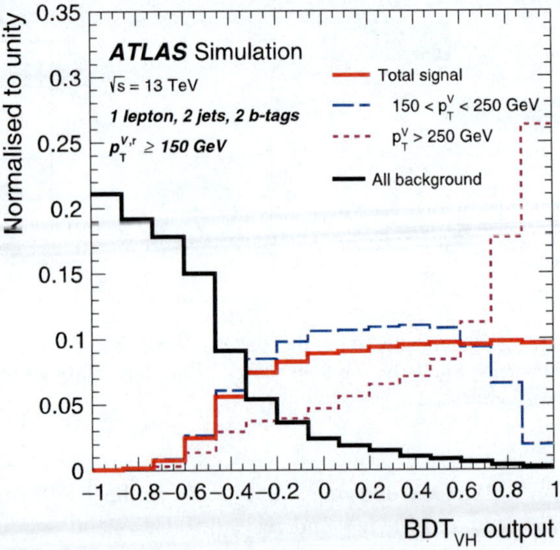


**Fig. 8.4** BDT distributions for different STXS regions in 1-lepton channel, in the 2jet reconstructed category. Image source: [6]

**Table 8.1** Fit results with uncertainties for the $VH$ simplified template cross section measurements times the $H \to b\bar{b}$ branching ratio (expressed in femtobarn), in the 5-POI (top five rows) and 3-POI (bottom three rows) schemes. Table source: [6]

| Measurement region | SM prediction | Result | Stat. unc. | Syst. unc. [fb] | | |
|---|---|---|---|---|---|---|
| ($\lvert y_H \rvert < 2.5, H \to b\bar{b}$) | [fb] | [fb] | [fb] | Th. sig. | Th. bkg. | Exp. |
| 5-POI scheme | | | | | | |
| $W \to \ell\nu$; $150 < p_T^V <$ 250Gev | $24.0 \pm 1.1$ | $20 \pm 25$ | $\pm 17$ | $\pm 2$ | $\pm 13$ | $\pm 9$ |
| $W \to \ell\nu$ $p_T^V >$ 250Gev | $71.1 \pm 0.3$ | $8.8 \pm 5.2$ | $\pm 4.4$ | $\pm 0.5$ | $\pm 2.5$ | $\pm 0.9$ |
| $Z \to \ell\ell, \nu\nu$ $75 < p_T^V <$ 150Gev | $50.6 \pm 4.1$ | $81 \pm 45$ | $\pm 35$ | $\pm 10$ | $\pm 21$ | $\pm 19$ |
| $Z \to \ell\ell, \nu\nu$ $150 < p_T^V <$ 250Gev | $18.8 \pm 2.4$ | $14 \pm 13$ | $\pm 11$ | $\pm 1$ | $\pm 6$ | $\pm 3$ |
| $Z \to \ell\ell, \nu\nu$ $p_T^V >$ 250Gev | $4.9 \pm 0.5$ | $8.5 \pm 4.0$ | $\pm 3.7$ | $\pm 0.8$ | $\pm 1.2$ | $\pm 0.6$ |
| 3-POI scheme | | | | | | |
| $W \to \ell\nu$; $p_T^V >$ 150Gev | $31.1 \pm 1.4$ | $35 \pm 14$ | $\pm 9$ | $\pm 2$ | $\pm 9$ | $\pm 4$ |
| $Z \to \ell\ell, \nu\nu$ $75 < p_T^V <$ 150Gev | $50.6 \pm 4.1$ | $81 \pm 45$ | $\pm 35$ | $\pm 10$ | $\pm 21$ | $\pm 19$ |
| $Z \to \ell\ell, \nu\nu$ $p_T^V >$ 150Gev | $23.7 \pm 3.0$ | $28.4 \pm 8.1$ | $\pm 6.4$ | $\pm 2.4$ | $\pm 3.6$ | $\pm 2.3$ |

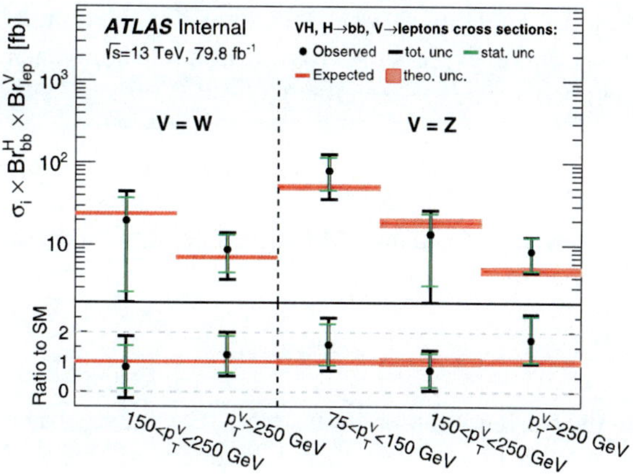


**Fig. 8.5** Measured $VH$ cross sections times the $H \to b\bar{b}$ branching ratio. Image source: [6]

larger if a finer splitting of the phase space in used. Among the systematic uncertainties, those related to the limited size of the MC background samples and their theoretical modelling are dominant. The uncertainties due to the signal modelling are found to be small, between 6 and 12%. The uncertainties on the SM predictions are larger for $ZH$ than $WH$, due to the limited precision of the theoretical calculations of $gg \to ZH$ process. The results for the 5-POI scheme shown in Fig. 8.5 agree with the SM prediction. With the 5-POI scheme, an anti-correlation of 40–60% is observed between the cross sections in the bins $150 < p_{\mathrm{T}}^{V} < 250$ GeV and $p_{\mathrm{T}}^{V} > 250$ GeV. This is due to the fact that the two regions correspond to the same analysis category.

## 8.4 Effective Lagrangian Interpretation

The definition of fiducial volumes allows the isolation of high $p_{\mathrm{T}}^{V}$ regions where the sensitivity to potential modifications of the Higgs boson couplings is expected to be enhanced. The STXS framework is particularly suitable for investigating potential deviations with the use of an effective field theory (EFT) approach, described in Sect. 2.5. A large number of operators enters the formulation of the effective Lagrangian, but a limited set of them affects the Higgs boson couplings. These operators can be constrained with Higgs physics measurements, and specifically with cross section measurements, by relating the STXSs to the operators. In this study, only dimension $D = 6$ operators are considered and the effective Lagrangian takes the form:

$$\mathscr{L}_{EFT} = \mathscr{L}_{SM} + \sum_{i} c_i^{(6)} \mathcal{O}_i^{(6)} / \Lambda^2, \tag{8.6}$$

where $\mathcal{O}_i^{(6)}$ represent dimension $D = 6$ operators, $c_i$ are coupling constants (the Wilson coefficients) and $\Lambda$ represents the energy scale of new physics. In order to establish a relationship between the STXSs and the effective Lagrangian operators, it is useful to express the cross section into components arising from SM, BSM and the interference between them:

$$\sigma_{EFT} = \sigma_{SM} + \sigma_{int} + \sigma_{BSM}. \tag{8.7}$$

The deviation from the SM can then be expressed as a function of the coefficients of the operators:

$$\frac{\sigma_{EFT}}{\sigma_{SM}} = 1 + \sum_i A_i \bar{c}_i + \sum_{ij} B_{ij} \bar{c}_i \bar{c}_j. \tag{8.8}$$

In the convention used in this work, the scale $\Lambda$ has been absorbed in the definition of the Wilson coefficients and the operators divided by $v^2$, where $v$ is the vacuum expectation value of the Higgs boson field, such that the coefficients considered $\bar{c}_i$ are dimentionless: $\bar{c}_i = c_i^{(6)} v^2/\Lambda^2$. In Eq. (8.8), $A_i$ and $B_{ij}$ define coefficients that can be derived through a leading order MADGRAPH generation, as documented in [8]. The non-SM contributions in Eq. (8.8) are given by $\frac{\sigma_{int}}{\sigma_{SM}} = \sum_i A_i \bar{c}_i$ and $\frac{\sigma_{BSM}}{\sigma_{SM}} = \sum_{ij} B_{ij} \bar{c}_i \bar{c}_j$, which are linear and quadratic terms in $\bar{c}_i$, respectively. The linear terms arise from the interference between SM and BSM, while the quadratic terms are the leading contribution from BSM only. In a similar way, the EFT operators can affect the coupling involved in the Higgs decay, thus modifying the corresponding BR. In order to parametrise the partial widths of the Higgs decays, an identical procedure as for the cross sections is used. For instance, the BR of the $H \rightarrow b\bar{b}$ process can be expressed as shown in Eq. (8.9):

$$\mathrm{BR}_{bb} = \frac{\Gamma_{bb}}{\Gamma_{All}} \approx \frac{\Gamma_{bb}^{SM}}{\Gamma_{All}^{SM}} \left[ 1 + \sum_i A_i^{bb} \bar{c}_i + \sum_{ij} B_{ij}^{bb} \bar{c}_i \bar{c}_j - \left( \sum_i A_i^{All} \bar{c}_i + \sum_{ij} B_{ij}^{All} \bar{c}_i \bar{c}_j \right) \right], \tag{8.9}$$

where $\Gamma_{bb}$ is the partial width of the $H \rightarrow b\bar{b}$ decay and $\Gamma_{All}$ is the total width.

## 8.5 $VH(H \rightarrow b\bar{b})$ EFT Parametrisation

A complete procedure for fitting STXS measurements to the coefficients of the effective Lagrangian operators has been developed by the LHC community, and documented in [8]. The document also reports the explicit dependencies of the $VH$ stage-1 STXS bins to the EFT operators based on leading order (LO) predictions, retrieved in the SILH formulation [7], introduced in Sect. 2.5. In order to convert the parameterisation from the stage-1 to the reduced stage-1, it is necessary to extrapolate the coefficients according to the relative combination of bins. In particular, the reduced stage-1 does not distinguish between events with or without additional jets with respect to the two $b$-tagged jets coming from the Higgs boson. The coefficients in the merged region are obtained by summing the original coefficients weighted by

**Table 8.2** The EFT operators and the corresponding HEL coefficients relevant for the $VH(H \to b\bar{b})$ measurements are listed. The Lagrangian term introduced by each operator and the affected vertices are also shown

| Operator | Lagrangian term | HEL coefficient | Vertices |
|---|---|---|---|
| $\mathcal{O}_{HW}$ | $i\,(D^{\mu}H)^{\dagger}\,\sigma^{a}\,(D^{\nu}H)\,W^{a}_{\mu\nu}$ | $\bar{c}_{HW} = \frac{m_W^2}{g}\frac{c_{HW}}{\Lambda^2}$ | $HWW/HZZ$ |
| $\mathcal{O}_{HB}$ | $i\,(D^{\mu}H)^{\dagger}\,(D^{\nu}H)\,B_{\mu\nu}$ | $\bar{c}_{HB} = \frac{m_W^2}{g'}\frac{c_{HB}}{\Lambda^2}$ | $H\gamma\gamma/HZZ$ |
| $\mathcal{O}_{W}$ | $i\left(H^{\dagger}\sigma^{a}D^{\mu}H\right)D^{\nu}W^{a}_{\mu\nu}$ | $\bar{c}_{W} = \frac{m_W^2}{g}\frac{c_W}{\Lambda^2}$ | $HWW/HZZ$ |
| $\mathcal{O}_{B}$ | $i\left(H^{\dagger}D^{\mu}H\right)D^{\nu}B_{\mu\nu}$ | $\bar{c}_{B} = \frac{m_W^2}{g'}\frac{c_B}{\Lambda^2}$ | $H\gamma\gamma/HZZ$ |
| $\mathcal{O}_{d}$ | $y_d\lvert H\rvert^2\bar{Q}_L H d_R + \text{h.c.}$ | $\bar{c}_{d} = v^2\frac{c_d}{\Lambda^2}$ | $Hbb$ |

the cross sections predicted by the SM. The coefficients for $ZH$ in the new region of $75 < p_{\mathrm{T}}^{V} < 150$ GeV are obtained by following the prescription in [8], which consists in generating MADGRAPH samples using an EFT model and assuming that the new operators act one at the time. Then, the cross section times the acceptance in the region of interest is evaluated for each sample with RIVET [9], a software that allows to compute fiducial physics objects and to analyse them, and compared with the same quantity obtained with the SM assumption. The parameterisations for the five STXS regions are then established.

The $VH(H \to b\bar{b})$ analysis is sensitive to the operators that introduce new Higgs boson interactions with the $W$ boson and the $Z$ boson (in the production process) and the $b$ quark (in the decay process). The operators considered are defined in the context of the *Higgs Effective Lagrangian* implementation (HEL) [10]. The constraints on the operators are obtained through a binned maximum likelihood fit performed on the kinematic regions identified for the 5-POI STXS. The cross sections are parametrised as a function of the dimensionless coefficients of the operators involved in the $VH$ processes. The following set of coefficients has been considered: $\bar{c}_{HW}, \bar{c}_{HB}, \bar{c}_{W} - \bar{c}_{B}$ and $\bar{c}_{d}$. The sum $\bar{c}_{W} + \bar{c}_{\mathrm{B}}$ is strongly constrained by precision electroweak data [12], therefore it is assumed to be zero in this work. $\bar{c}_{HW}$ and $\bar{c}_{W}$ are associated to the operators affecting the interaction of the Higgs boson with the field that generates the $HWW$ and $HZZ$ vertices, while $\bar{c}_{HB}$ and $\bar{c}_{B}$ introduce contributions to the $HZZ$ vertex. The $\bar{c}_{d}$ coefficient affects the coupling between the Higgs boson and bottom quarks through the modification of the partial width $\Gamma_H^{bb}$ and the total width $\Gamma_H^{All}$ of the Higgs boson. The operators and the corresponding coefficients are listed in Table 8.2, where $g$ and $g'$ represent the SU(2) and U(1) SM gauge couplings, and $v$ is the vacuum expectation value of the Higgs boson field.

The equations used for the parametrisation of the cross sections, which include both linear ($\sum_i A_i\bar{c}_i$) and quadratic terms ($\sum_{ij} B_{ij}\bar{c}_i\bar{c}_j$), are reported in Table 8.3.

Analogously, Table 8.4 summarises the dependences of the partial and total widths from the coefficients. The coefficients that affect the $VH$ production cross sections, such as $\bar{c}_{HW}, \bar{c}_{HB}, \bar{c}_{W}$ and $\bar{c}_{B}$, also enter the $H \to b\bar{b}$ BR parametrisation, with small contributions.

**Table 8.3** The cross section parameterisations are listed for each $VH(H \to b\bar{b})$ STXS region. The contribution from the interference between SM and BSM ($\sum_i A_i \bar{c}_i$) and the pure BSM terms ($\sum_{ij} B_{ij} \bar{c}_i \bar{c}_j$), are reported separately

| Cross section region | $\sum_i A_i \bar{c}_i$ |
|---|---|
| $q\bar{q} \to Hl\nu$ ($150 \leq p_T^V \leq 250$) GeV | $50\bar{c}_{HW} + 74\bar{c}_W$ |
| $q\bar{q} \to Hl\nu$ ($p_T^V \geq 250$) GeV | $170\bar{c}_{HW} + 200\bar{c}_W$ |
| $q\bar{q} \to Hll$ ($75 \leq p_T^V \leq 150$) GeV | $13\bar{c}_{HW} + 38\bar{c}_W + 3.9\bar{c}_{HB} + 10.5\bar{c}_B$ |
| $q\bar{q} \to Hll$ ($150 \leq p_T^V \leq 250$) GeV | $37\bar{c}_{HW} + 61\bar{c}_W + 11\bar{c}_{HB} + 18\bar{c}_B$ |
| $q\bar{q} \to Hll$ ($p_T^V \geq 250$) GeV | $130\bar{c}_{HW} + 150\bar{c}_W + 38\bar{c}_{HB} + 46\bar{c}_B$ |
| Cross section region | $\sum_{ij} B_{ij} \bar{c}_i \bar{c}_j$ |
| $q\bar{q} \to Hl\nu$ ($150 \leq p_T^V \leq 250$) GeV | $839\bar{c}_{HW}^2 + 1555\bar{c}_W^2 + \bar{c}_{HW}(900\bar{c}_W)$ |
| $q\bar{q} \to Hl\nu$ ($p_T^V \geq 250$) GeV | $14000\bar{c}_{HW}^2 + 16000\bar{c}_W^2 + \bar{c}_{HW}(30000\bar{c}_W)$ |
| $q\bar{q} \to Hll$ ($75 \leq p_T^V \leq 150$) GeV | $85\bar{c}_{HW}^2 + 400\bar{c}_W^2 + 8\bar{c}_{HB}^2 + 35\bar{c}_B^2$ |
| | $+\bar{c}_{HW}(150\bar{c}_W + 20\bar{c}_{HB} + 42\bar{c}_B)$ |
| | $+\bar{c}_{HB}(44\bar{c}_W + 12\bar{c}_B) + \bar{c}_W(140\bar{c}_B)$ |
| $q\bar{q} \to Hll$ ($150 \leq p_T^V \leq 250$) GeV | $462\bar{c}_{HW}^2 + 982\bar{c}_W^2 + 41\bar{c}_{HB}^2 + 86\bar{c}_B^2$ |
| | $+\bar{c}_{HW}(1255\bar{c}_W + 277\bar{c}_{HB} + 358\bar{c}_B)$ |
| | $+\bar{c}_{HB}(373\bar{c}_W + 105\bar{c}_B) + \bar{c}_W(587\bar{c}_B)$ |
| $q\bar{q} \to Hll$ ($p_T^V \geq 250$) GeV | $8000\bar{c}_{HW}^2 + 9600\bar{c}_W^2 + 720\bar{c}_{HB}^2 + 850\bar{c}_B^2$ |
| | $+\bar{c}_{HW}(17000\bar{c}_W + 4800\bar{c}_{HB} + 5100\bar{c}_B)$ |
| | $+\bar{c}_{HB}(5100\bar{c}_W + 1500\bar{c}_B) + \bar{c}_W(5700\bar{c}_B)$ |

**Table 8.4** Parametrisations for the interference between SM and BSM amplitudes and pure BSM contribution for the $H \to b\bar{b}$ Branching Ratio

| Partial width | $\sum_i A_i \bar{c}_i$ | $\sum_{ij} B_{ij} \bar{c}_i \bar{c}_j$ |
|---|---|---|
| $H \to b\bar{b}$ | $3\bar{c}_d$ | $2.3\bar{c}_d^2$ |
| $H \to$ All | $2.3\bar{c}_d + 0.37\bar{c}_{HW} + 1.0\bar{c}_W$ $+0.0079\bar{c}_{HB} + 0.023\bar{c}_B$ | $1.7\bar{c}_d^2 + 4.7\bar{c}_{HW}^2$ $+2.6\bar{c}_W^2 + 4.3\bar{c}_{HB}^2$ |

These contributions have been taken into account in the parametrisation used for the fits.

As an example, the curves that correspond to the EFT parametrisation of the cross sections induced by the $\bar{c}_W - \bar{c}_B$ coefficient is shown in Fig. 8.6.

The curves illustrate the effect on the cross section for each kinematic region, for the combined linear and quadratic contributions (solid lines), and for linear contribution only (dashed lines).

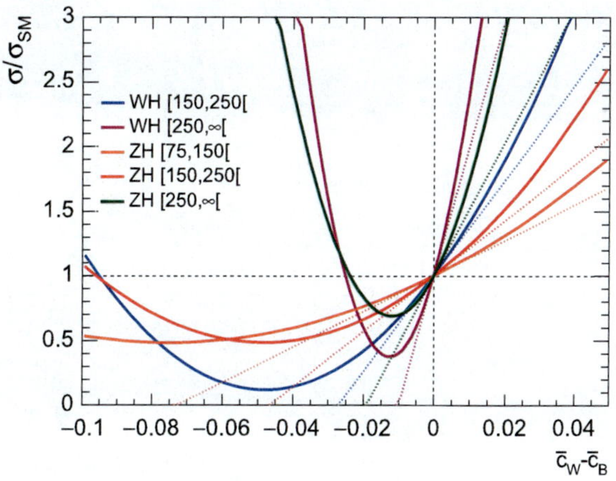


**Fig. 8.6** Cross section variations for each kinematic region introduced by the $\bar{c}_W - \bar{c}_B$ coefficient. The combined linear and quadratic contributions are illustrated by solid lines, while linear contributions only are illustrated by dashed lines

## 8.6 EFT Fit Results

Simultaneous maximum likelihood fits to the five STXS bins are performed to find the constraints for the EFT coefficients. A one-dimensional fit is performed for each of the EFT parameters considered, while the other coefficients are kept constant at zero, as predicted by the SM. The parametrisation considered includes both linear and quadratic contributions. In Fig. 8.7 the likelihood profile is shown for each coefficient. The likelihood presents a typical shape with two minima, due to the parabolic parametrisation used.

The fits are repeated by using a parametrisation which only includes linear contributions. The likelihood profile obtained is shown in Fig. 8.8 for each parameter and is compared to the previous result. The likelihood function has now a single minimum, since the parametrisation used is monotonic with respect to each coefficient.

The expected and observed intervals for the fitted EFT coefficients are reported in Table 8.5 with 68% and 95% of confidence level (CL). Using both the linear and quadratic parametrisation, the parameters $\bar{c}_{HW}$ and $\bar{c}_W - \bar{c}_B$ are constrained within a few percent at 95% CL, while the constraint on $\bar{c}_{HB}$ is about five times worse. In fact, since $\bar{c}_{HB}$ only affects the $ZH$ production, the analysis is less sensitive to the measurement of this coefficient. Similarly, the constraint on $\bar{c}_d$ is not very powerful since this coefficient is expected to modify only the decay coupling. The constraints are about 50% stronger when considering only the linear parametrisation.

The coefficient $\bar{c}_d$ has been constrained for the first time in this work, while the set of coefficients involved in the $VH$ production was considered in previous publications. An ATLAS publication [11] reports constraints obtained in the combined $H \to ZZ^* \to 4\ell$ and $H \to \gamma\gamma$ channels using 36.1 $fb^{-1}$ of data. Although the implementation used is the same as in this thesis, the mentioned work performs a simultaneous fit on six coefficients, therefore the comparison of the results is not straightforward. Another publication [12] describes a global fit to Higgs, Diboson

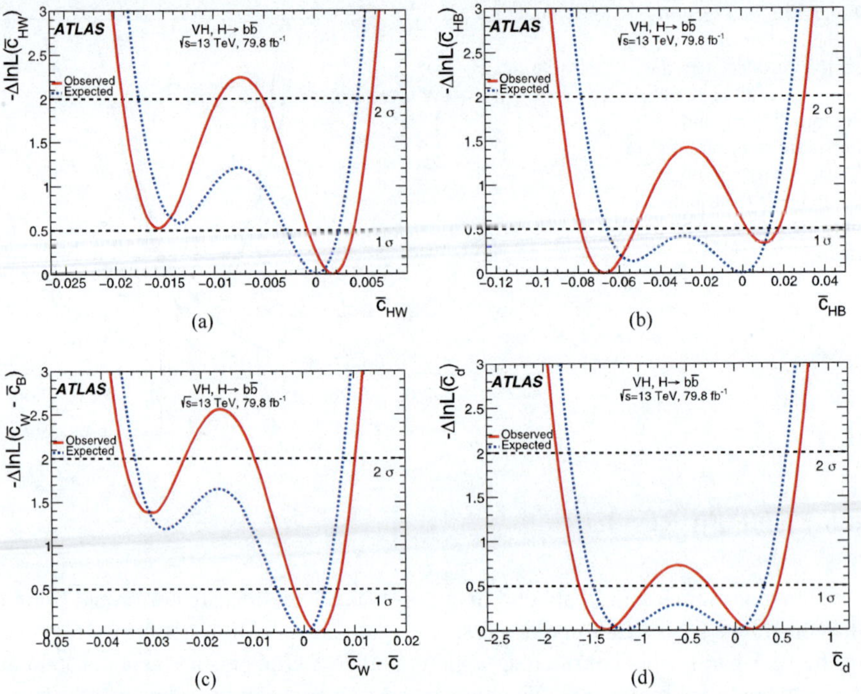


**Fig. 8.7** Expected (blue) and observed (red) profile likelihoods of the fitted EFT coefficients. The parametrisation used includes both linear and quadratic contributions. Image source: [6]

and EW data and uses only linear parametrisation for the coefficients constrained. The sensitivity reached by the $VH(H \to b\bar{b})$ analysis to anomalous couplings of the Higgs boson would certainly bring a great contribution in future global measurements.

The expected effect of the EFT parameters on the SM cross sections for the STXS bins is shown in Fig. 8.9. The variations, expressed in terms of $\sigma_i \times \mathrm{BR}_{bb}/\sigma_i^{\mathrm{SM}} \times \mathrm{BR}_{bb}^{SM}$, are obtained by assuming for each parameter the value of the upper limit of the corresponding interval measured with 95% of CL. As expected, the $\bar{c}_{HW}$ and $\bar{c}_W$ parameters affect all bins, while the $\bar{c}_{HB}$, $\bar{c}_B$ parameters only affect bins associated to the $ZH$ production mode. The $\bar{c}_d$ parameter has the same impact across the bins, because it contributes to the $H \to b\bar{b}$ BR.

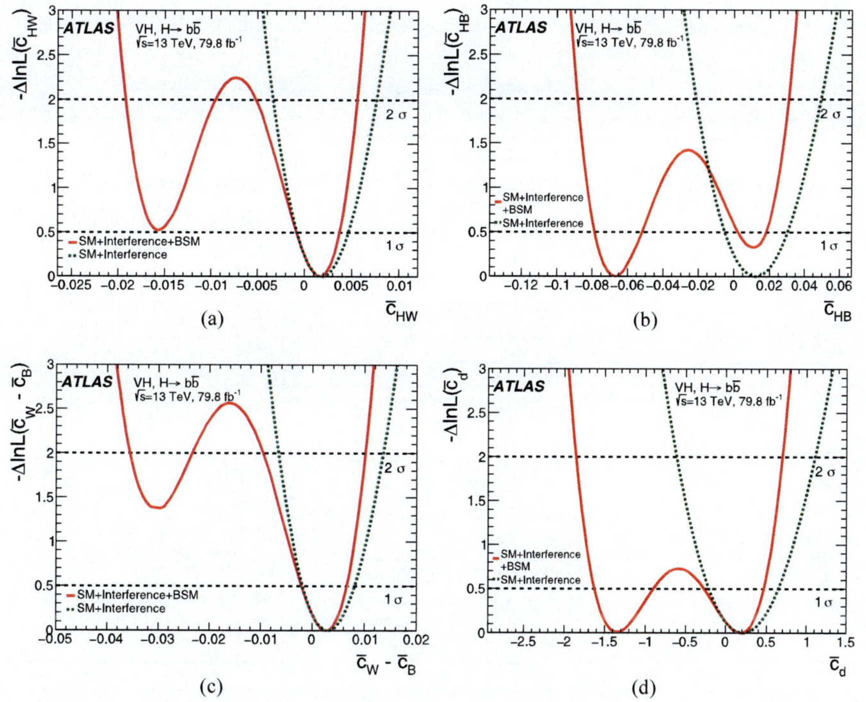


**Fig. 8.8** Observed profile likelihoods of the fitted EFT parameters. Comparison between the result obtained with the only linear parametrisation (green) and the result with the full (linear+quadratic) one (red). Image source: [6]

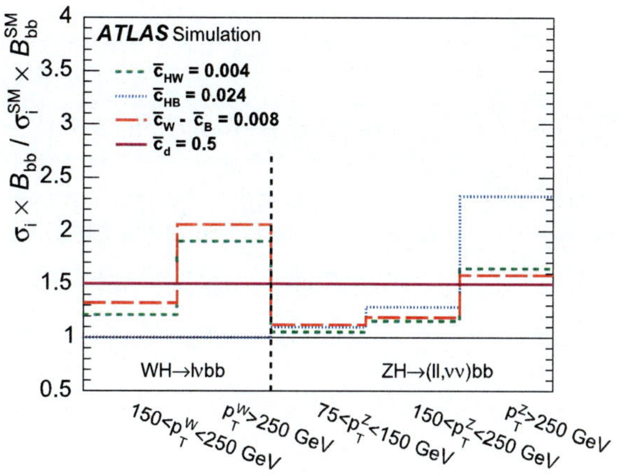


**Fig. 8.9** Variations on the SM cross sections induced by the EFT coefficients considered in the $VH(H \to b\bar{b})$ measurements. Image source: [6]

**Table 8.5** The expected and observed 68 and 95% confidence intervals for the EFT coefficients $\bar{c}_{HW}, \bar{c}_{HB}, \bar{c}_W - \bar{c}_B$ and $\bar{c}_d$. Table source: [6]

| Coefficient | Expected interval | Observed interval |
|---|---|---|
| Results at 68% confidence level | | |
| $\bar{c}_{HW}$ | [–0.003, 0.002] | [–0.001, 0.004] |
| (interference only | [–0.002, 0.003] | [–0.001, 0.005]) |
| $\bar{c}_{HB}$ | [–0.006, 0.013] | [–0.078, –0.055] ⋃ [0.005, 0.019] |
| (interference only | [–0.016, 0.016] | [–0.005, 0.030]) |
| $\bar{c}_W - \bar{c}_B$ | [–0.006, 0.005] | [–0.002, 0.007] |
| (interference only | [–0.005, 0.005] | [–0.002, 0.008]) |
| $\bar{c}_D$ | [–1.5, 0.3] | [–1.6, –0.9] ⋃ [–0.3, 0.4] |
| (interference only | [–0.4, 0.4] | [–0.2, 0.65]) |
| Results at 95% confidence level | | |
| $\bar{c}_{HW}$ | [–0.018, 0.004] | [–0.019, –0.010] ⋃ [–0.005, 0.006] |
| (interference only | [–0.005, 0.005] | [–0.003, 0.008]) |
| $\bar{c}_{HB}$ | [–0.078, 0.024] | [–0.090, –0.032] |
| (interference only | [–0.033, 0.033] | [–0.022, 0.049]) |
| $\bar{c}_W - \bar{c}_B$ | [–0.034, 0.008] | [–0.036, 0.024] ⋃ [–0.009, 0.010] |
| (interference only | [–0.009, 0.010] | [–0.006, 0.014]) |
| $\bar{c}_D$ | [–1.7, 0.5] | [–1.9, 0.7] |
| (interference only | [–0.8, 0.8] | [–0.6, 1.1]) |

## 8.7 Considerations on the EFT Interpretation

The EFT interpretation work performed in the context of the $VH(H \to b\bar{b})$ cross section measurements represents a proof of concept for the methodology necessary for a future global fit to precision data. Several arguments have been discussed in order to construct an appropriate approach to the EFT interpretation for $VH(H \to b\bar{b})$. In this section, the most important considerations are described.

### 8.7.1 $gg \to ZH$ Contribution

From a theoretical point of view, the $VH$ production can be split into a Drell-Yan component and a gluon-gluon component. As shown in the Feynman diagrams of Fig. 8.10, the $gg \to ZH$ process appears only at NNLO, through a top quark triangular loop or a box loop.

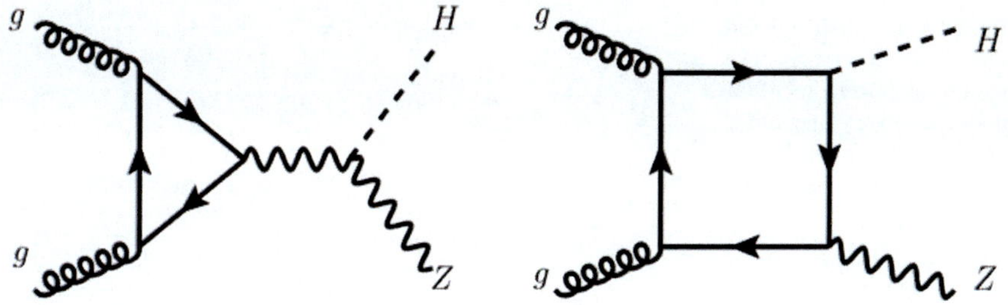


**Fig. 8.10** Feynman diagram topologies contributing to the $gg \to ZH$ production

The standard EFT approach is to fit using the LO relative correction to the SM from dimension-6 operators. The EFT samples are produced with a LO MADGRAPH generation and do not contain loop-induced diagrams. Modifications of the $gg \to ZH$ production cross section are introduced by higher dimensions and so they are not included in the correction in the current approach. In the $VH(H \to b\bar{b})$ EFT interpretation, the $gg \to ZH$ cross section is fixed to the SM value. Nevertheless, even if contributing at a higher order, the $gg \to ZH$ process is enhanced by the large gluon-gluon luminosity at the LHC (due to the large values of the PDFs) and it can contribute up to ~30% on a single STXS bin. Additionally, the gluon components becomes more important for high $p_{\mathrm{T}}^{V}$, due to the top quark loop threshold [13]. An identical EFT parametrisation as $qq \to ZH$ (which contributes at LO) has been assumed to perform a fit to be compared with the standard setup where $gg \to ZH$ is assumed not to be affected by the EFT operators. The fit provides stronger limits. However, the $qq \to ZH$ parametrisation is not appropriate for possible $gg \to ZH$ modifications. The standard approach of assuming the SM value for the corresponding cross section has been therefore used.

### 8.7.2 EFT in 3-POI Scheme

The 5-POI results have been compared with the fits obtained on the 3-POI scheme. The parametrisation for the regions resulting from merging two adjacent regions of the 5-POI scheme is obtained by summing the corresponding coefficients weighted by the relative cross section. The 5-POI fit results in stronger constraints, approximately improving the $\bar{c}_{HW}$ sensitivity by a factor two, as shown in Fig. 8.11. The better constraining power is due to the larger contribution of EFT operators at higher values of $p_{\mathrm{T}}^{V}$.

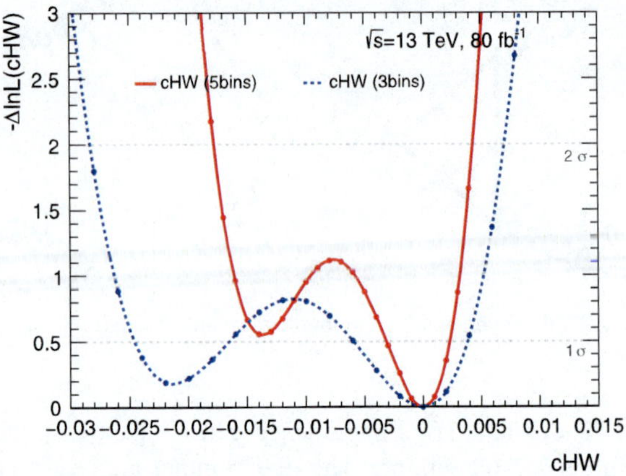


**Fig. 8.11** Profile likelihood for the $\bar{c}_{HW}$ coefficient. The main results from the 5-POI model (red) are compared to the results of the 3-POI model parametrisation (blue)

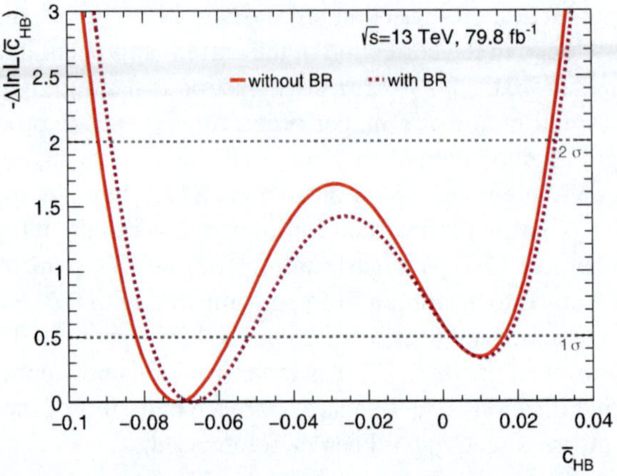


**Fig. 8.12** Profile likelihood fit comparison for $\bar{c}_{HB}$ coefficient in case the BR parametrisation is included (red) or not included (purple) in the EFT parametrisation

### 8.7.3 Branching Ratio Contribution

As shown in Table 8.4, the EFT coefficients affecting the $VH$ production, also contribute to the BR modification. These contributions are in general small but in some cases not negligible, especially when considering $\bar{c}_{HW}$ and $\bar{c}_{HB}$ coefficients. As an example, in Fig. 8.12 the profile likelihood fits for the $\bar{c}_{HB}$ coefficient are shown in case the BR parametrisation is included (red) or not included (purple) in the parametrisation.

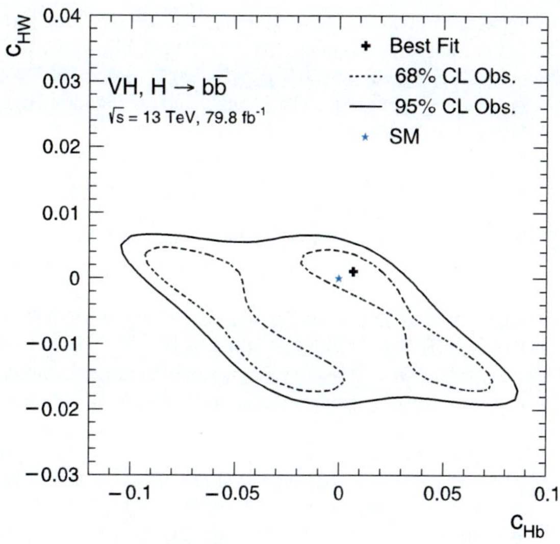


**Fig. 8.13** Contour plots at 68% (dashed) and 95% (full) CL from the simultaneous fit of two EFT coefficients ($\bar{c}_{HW}$ and $\bar{c}_{HB}$) to the observed data

### *8.7.4 Multi-dimensional Fits*

Simultaneous fits with two parameters have been performed for the $\bar{c}_{HW}$ and $\bar{c}_{HB}$ coefficients, using both linear and quadratic parametrisation, as shown in Fig. 8.13. The correlations between the two parameters are found to be very strong ($\sim$70%), therefore it has been decided to publish only one-dimentional fit results for the constraint of the EFT coefficients relevant to the $VH(H \to b\bar{b})$ analysis.

## References


1. Boggia M et al (2018) The HiggsTools handbook: a beginners guide to decoding the Higgs sector. J Phys G45(6):065004. https://doi.org/10.1088/1361-6471/aab812. arXiv:1711.09875 [hep-ph]
2. Boudjema F et al (2013) On the presentation of the LHC Higgs results. In: Workshop on Likelihoods for the LHC Searches Geneva, Switzerland, January 21-23, 2013. arXiv:1307.5865 [hep-ph]
3. de Florian D et al (2016) Handbook of LHC Higgs cross sections: 4. Deciphering the nature of the Higgs sector. In: (2016). https://doi.org/10.23731/CYRM-2017-002. arXiv:1610.07922 [hep-ph]
4. Andersen JR et al (2015) Les Houches 2015: Physics at TeV Colliders Standard Model Working Group Report. In: 9th Les Houches Workshop on Physics at TeV Colliders (PhysTeV 2015) Les Houches, France, June 1-19, 2015. 2016. arXiv: 1605.04692 [hep-ph]. url: http://lss.fnal.gov/archive/2016/conf/fermilab-conf-16-175-ppd-t.pdf
5. Mimasu K, Sanz V, Williams C (2016) Higher order QCD predictions for Associated Higgs production with anomalous couplings to gauge bosons. JHEP 08:039. https://doi.org/10.1007/JHEP08(2016)039. arXiv:1512.02572 [hep-ph]

6. Aaboud M et al (2019) Measurement of VH, H → bb production as a function of the vector-boson transverse momentum in 13 TeV pp collisions with the ATLAS detector. JHEP 05:141. https://doi.org/10.1007/JHEP05(2019)141. arXiv:1903.04618 [hep-ex]
7. Giudice GF et al (2007) The strongly-interacting light Higgs. JHEP 06:045. https://doi.org/10.1088/1126-6708/2007/06/045. arXiv:hep-ph/0703164 [hep-ph]
8. Hays C, Gonzalez VS, Zemaityte G (2019) Constraining EFT parameters using simpli?ed template cross sections. Technical report. LHCHXSWG-2019-004. Geneva: CERN, May 2019. https://cds.cern.ch/record/2673969
9. Bierlich C et al (2020) Robust independent validation of experiment and theory: rivet version 3. SciPost Phys 8:026. https://doi.org/10.21468/SciPostPhys.8.2.026. arXiv:1912.05451 [hep-ph]
10. Alloul A, Fuks B, Sanz V (2014) Phenomenology of the Higgs Effective Lagrangian via FEYNRULES. JHEP 04:110. https://doi.org/10.1007/JHEP04(2014)110. arXiv:1310.5150 [hep-ph]
11. Constraints on an effective Lagrangian from the combined H → ZZ∗ → 4$\ell$ and H → $\gamma\gamma$ channels using 36.1 fb$^{-1}$ of $\sqrt{s}$ = 13 TeV pp collision data collected with the ATLAS detector. Technical report. ATL-PHYS-PUB-2017-018. Geneva: CERN, Nov. 2017. https://cds.cern.ch/record/2293084
12. Ellis J et al (2018) Updated global SMEFT fit to Higgs, Diboson and electroweak data. JHEP 06:146. https://doi.org/10.1007/JHEP06(2018)146. arXiv:1803.03252 [hep-ph]
13. Englert C, McCullough M, Spannowsky M (2014) Gluon-initiated associated production boosts Higgs physics. Phys Rev D89(1):013013. https://doi.org/10.1103/PhysRevD.89.013013. arXiv:1310.4828 [hep-ph]

# Chapter 9
# Conclusions and Future Prospects

The discovery of the Higgs boson represents an outstanding success of particle physics. It opened a window for testing the scalar sector of the Standard Model and its possible extensions. All the measurements performed during Run-1 and Run-2 are found to be compatible with the Standard Model.

The Run-2 analysis that enabled the observation of the Higgs decay into a pair of bottom quarks, as well as the associated production with a vector boson, has been presented in this thesis. The $VH(H \rightarrow b\bar{b})$ analysis is performed on a dataset corresponding to an integrated luminosity of 79.8 $fb^{-1}$, collected by the ATLAS experiment at a centre-of-mass energy of 13 TeV. The leptonic decays of the vector boson allow for an efficient triggering and background rejection. The analysis is divided into three channels, depending on the number of leptons in the final state, and the event selection is optimised in each channel. The analysis greatly benefits from the discrimination between signal and background events provided by a multivariate analysis based on a Boosted Decision Tree. After training the Boosted Decision Tree, a maximum likelihood fit is performed to extract the signal strength and normalisation of the main backgrounds. An excess of events over the expected Standard Model background is found with an observed (expected) significance of 4.9 (4.3) standard deviation. A combination with results from other searches for the $H \rightarrow b\bar{b}$ process is performed, providing an observed (expected) significance of 5.4 (5.5). The corresponding ratio between the signal yield and the SM expectation is $1.01 \pm 0.12(\text{stat.})^{+0.16}_{-0.15}(\text{syst.})$.

This thesis has also reported the evolution of the $VH(H \rightarrow b\bar{b})$ analysis as a measurement of the $VH$ production cross sections times the $H \rightarrow b\bar{b}$ branching ratio as a function of the vector boson transverse momentum. The measurement is performed in exclusive regions of phase space, using the Simplified Template Cross Section framework. All measurements are found to be in agreement with the Standard Model predictions and are used to search for possible deviations, using an effective field theory approach, based on anomalous couplings of the Higgs boson. An effective



C. Tosciri, *Higgs Boson Decays into a Pair of Bottom Quarks*, Springer Theses,
https://doi.org/10.1007/978-3-030-87938-9_9

Lagrangian, sensitive to dimension-six operators that could potentially modify the Higgs boson couplings to the electroweak gauge bosons, has been constrained. The effective field theory interpretation of the $VH(H \rightarrow b\bar{b})$ cross sections represents a preparatory ground work for further implementations towards the first global EFT fit which is being prepared by the ATLAS Collaboration. The global fit will include measurements from other Higgs boson decay channels and electroweak processes.

Following the release of the results contained in this thesis, the analysis team updated the $VH(H \rightarrow b\bar{b})$ measurements with the full dataset collected in Run-2, corresponding to an integrated luminosity of 139 $fb^{-1}$. The new analysis [1] profits from an increased statistics as well as from an improved analysis strategy.

The successful data collection of Run-2 allowed the experiments at the LHC to improve the methodologies to study the properties of the Higgs boson. Many projections on the HL-LHC scenario predict very high precision for Higgs measurements, which will be dominated by theoretical uncertainties. The couplings of the Higgs to the vector bosons and to the third generation fermions will be measured to the percent level. In particular, the studies performed by extrapolating the $VH(H \rightarrow b\bar{b})$ measurements of Run-2 to the HL-LHC scenario predict a precision of $\sim 5\%$ for the coupling of the Higgs boson to $b$-quarks [2, 3].

The HL-LHC will also enable the measurement of the tri-linear coupling of the Higgs boson to itself, with a precision of 50%. This is an important parameter of the electroweak sector, directly related to the shape of the Higgs potential, that could show sizeable deviations from the SM prediction. The direct way to measure Higgs self-interaction is through the production of Higgs pairs via gluon-gluon fusion. Due to the small cross sections (about three orders of magnitude smaller than the single Higgs production), the search for these processes is very challenging. The most sensitive channels are provided by decays with $b$-quarks in the final state, like $b\bar{b}b\bar{b}$, $b\bar{b}\gamma\gamma$ and $b\bar{b}\tau\tau$. The $b\bar{b}b\bar{b}$ final state has the advantage of a large BR ($\sim$ 33%), but the overwhelming QCD multi-jet background makes the search of Higgs pairs particularly challenging in this final state. As shown in the phenomenological studies documented in Ref. [4, 5], the application of classification algorithms based on machine learning can help to discriminate the Higgs pair signal from the QCD background, and a regression technique can also be exploited as an additional tool for the improvement of the classification. The di-Higgs measurements will greatly benefit from the strategies and studies adopted in previous analyses, like the $VH(H \rightarrow b\bar{b})$ analysis presented in this work [2].

A vast physics program is expected at the ATLAS experiment in the next decades. As the LHC luminosity will increase, the computing resources will not cope with the increasing simulation demand. Advanced techniques will be vital to provide fast and accurate simulations. This thesis presented a novel method, based on similarity search techniques, for the fast simulation of the forward calorimeter response in ATLAS. The new approach provides optimal results in terms of detector resolution response and a gain of about 70% in computational time, if compared with a standard full simulation.

# References


1. Aad G et al (2021) Measurements of W H and ZH production in the H $\to b\bar{b}$ decay channel in pp collisions at 13 TeV with the ATLAS detector. Eur Phys J C 81.2:178. https://doi.org/10.1140/epjc/s10052-020-08677-2. arXiv:2007.02873 [hep-ex]
2. Ellis RK et al (2019) Physics briefing book. arXiv:1910.11775 [hep-ex]
3. Projections for measurements of Higgs boson cross sections, branching ratios, coupling parameters and mass with the ATLAS detector at the HL-LHC. Technical report ATL-PHYS- PUB-2018-054. Geneva: CERN, Dec 2018. https://cds.cern.ch/record/2652762
4. AMVA4NewPhysics authors. D1.1 Multivariate Analysis Methods for Higgs Boson Searches at the Large Hadron Collider. https://amva4newphysics.wordpress.com/deliverables/
5. Tosciri C, Bortoletto D (2017) Classification studies for Higgs pair production in the resolved $b\bar{b}b\bar{b}$ final state. Technical report ATL-COM-PHYS-2017-1628. Geneva: CERN, Oct 2017. https://cds.cern.ch/record/2290881